T0122837

Beaked Whales

Beaked Whales

A COMPLETE GUIDE TO THEIR BIOLOGY AND CONSERVATION

Richard Ellis   James G. Mead

Illustrations by Richard Ellis

JOHNS HOPKINS UNIVERSITY PRESS   BALTIMORE

 Published 2017
Printed in China on acid-free paper
9 8 7 6 5 4 3 2 1

Johns Hopkins University Press
2715 North Charles Street
Baltimore, Maryland 21218-4363
www.press.jhu.edu

Library of Congress Cataloging-in-Publication Data

Names: Ellis, Richard, 1938– | Mead, James G.
Title: Beaked whales : a complete guide to their biology and conservation / Richard Ellis, James G. Mead.
Description: Baltimore : Johns Hopkins University Press, 2017. | Includes bibliographical references and index.
Identifiers: LCCN 2016019646| ISBN 9781421421827 (hardcover : alk. paper) | ISBN 1421421828 (hardcover : alk. paper) | ISBN 9781421421834 (electronic) | ISBN 1421421836 (electronic)
Subjects: LCSH: Beaked whales. | Toothed whales. | Whales.
Classification: LCC QL737.C438 E45 2017 | DDC 599.5/45—dc23
LC record available at https://lccn.loc.gov/2016019646

A catalog record for this book is available from the British Library.

Illustrations by Richard Ellis unless otherwise credited.

Special discounts are available for bulk purchases of this book. For more information, please contact Special Sales at 410-516-6936 or specialsales@press.jhu.edu.

Johns Hopkins University Press uses environmentally friendly book materials, including recycled text paper that is composed of at least 30 percent post-consumer waste, whenever possible.

Contents

ABOUT BEAKED WHALES

Preface

In 1872, William Henry Flower, director of Britain's Natural History Museum, made the following observations about beaked whales: "Their very presence in the ocean seems to pass unnoticed and unsuspected by voyagers, and even by those whose special occupation is the pursuit and capture of various better known and more abundant cetaceans until one of the accidental occurrences just alluded to reveals the existence of animal life of considerable magnitude (for they range between fifteen and twenty feet in length) and at least sufficiently numerous to maintain the continuity of the race."

Our knowledge improved only slightly as the twentieth century dawned. In 1910, Frederick True of the Smithsonian's United States National Museum wrote that "the beaked whales . . . are among the rarest of cetaceans," and he was able to locate only about a hundred specimens in the museum collections of the world. The beaked whales are still, by all criteria, the least known of all the large animals in the world, even though they live in all the oceans and are often large and imposing, the largest reaching a length of 42 feet.

We do not even know where most beaked whales live because most of our information comes from the examination of specimens that washed up on various beaches. It might be said that we know more about where they die than where they live. We are ignorant of where or when they breed, and we are not really sure how many different species there are. In *Wildlife in America* (1959), Peter Matthiessen wrote, "Some species of small beaked whales, an ancient group which may be passing slowly from existence, appear rarely in North American waters. There are only a few scattered records, on the Atlantic coast, of the Sowerby and the Gervais whales, and of the Stejneger's whale in the Pacific, and the habits of these mysterious creatures are virtually unknown."

As famed biologist E. O. Wilson noted in his sweeping discussion of the diversity of life, "Not even the Cetacea, containing the largest animals on earth, the whales and porpoises, is fully known . . . the beaked whales and porpoises have continued to yield new species at an average rate of one a decade during the twentieth century." Of the fourteen cetaceans discovered since 1908, nine are beaked whales.

Beaked whales were once thought of as poorly known, little-studied, deep-water "mysterious creatures." Today they are used in high-tech cetological research. Researchers acknowledge them as having the most sophisticated echolocators and being the deepest-diving whales and most accomplished suction feeders. Unfortunately, they are also the most sonar threatened of all cetaceans, suffering great damage when exposed to military sonar. Doors to the heretofore hidden world of beaked whales are being opened by scientists who are tracking them, photographing them, listening to them, and reclassifying them.

The typical beaked whale has a tapered, fusiform body, wide tail flukes that are not notched like those of most other cetaceans, small flippers that can be tucked into shaded "pockets," and some sort of an elongated snout that is a function of the elongated rostrum of the skull, which accounts for the name "beaked whale." (Naturally, there are exceptions.) Some of the species are known by much more descriptive common names, such as "goosebeak whale," "strap-toothed whale," or "ginkgo-toothed whale," and one species is known in New Zealand as the "scamperdown whale," but nobody seems to know why. There is a crescent-shaped blowhole with the horns pointing forward and a small dorsal fin that is usually located closer to the tail flukes than to the head. All members of the family have two throat grooves that converge toward the beak and relatively small eyes, located back and away from the gape of the mouth.

The teeth of beaked whales are among their most unusual features. In most species, only the males have erupted teeth, and there are usually only two of them located around the midpoint or the apex of the lower jaw. Females and juveniles of most species have teeth, but they remain unerupted, embedded in the mandible. Because they are suction feeders, beaked whales do not use their teeth for feeding, but rather for dominance battles between adult males. They are almost exclusively inhabitants of deep, offshore waters, and most of the evidence for their existence has come from bones picked up on the beach or from stranded or beached specimens.

Gol'din (2013) has come up with a novel and reasonable theory about the functional significance of the dense rostral bones in many of the males of beaked whale species. Because the differences in density of the rostrum would be sensed by an echolocating animal, it would be obvious to both sexes and would function much the same way antlers do in deer. The males that had strongly ossified rostra would be comparable to a buck with a magnificent rack and triumph in battles for mating rights.

For millennia, most beaked whales led their "specialized cetacean lives" unaffected by human intervention. Boats did not cross their paths; ship-generated noises did not reach them; whalers did not chase them down and kill them (Baird's beaked whale of the North Pacific and the northern bottlenose whale of the North Atlantic were exceptions). They swam unseen and unmolested in the vast blue depths of the open ocean, singing their clicks and whistles into the abyss, interacting only with family members, other families, other animals. Now they are threatened by sonar.

Unlike many other whale species, beaked whales do not venture into shallow water unless they intend to die there; otherwise, they spend their entire lives in the offshore depths, diving toward the bottom to capture the cephalopods that comprise almost their entire menu. Like dolphins, they echolocate; like the sperm whale, the narwhal, and the beluga, they suck up their food without using their teeth; and like the narwhal, only the males, with one exception, have visible teeth.

This brief introduction does the beaked whales little justice, but we hope the remainder of this book shows you why they have fascinated so many marine mammal biologists.

Acknowledgments

The scientific world has already acknowledged many of the great people involved with beaked whales. Fourteen species include the name of a *person*, either the one who described the species in the first place, or someone who was honored by having a whale named for him. To them we first acknowledge our gratitude: Roy Chapman Andrews (1884–1960); Maurice Arnoux (who found the skull of his namesake whale on a New Zealand beach); Spencer Fullerton Baird (1823–1887); Henri Marie Ducrotay de Blainville (1777–1850); Georges Cuvier (1869–1832); Paul F. L. Gervais (1816–1879); John Edward Gray (1800–1875), Sir James Hector (1835–1907); Carl Hubbs (1894–1979); Albert Heber Longman (1880–1954); James Sowerby (1757–1822); Leonhard Stejneger (1851–1943); George Shepherd (who collected the first specimens of *Tasmacetus shepherdi*); and Frederick W. True (1858–1914) and William F. Perrin (b. 1938), who is quite alive.

We hardly know where to begin when it comes to expressing our remaining gratitude. For decades we have learned so much from hundreds of nature's observers—be they people who sent us a stranded whale, told us of an encounter, or shared a scientific article. There are so many people who helped us along the way that we are certain to omit some of them. We begin by thanking Ed Asper, Don Ljungblad, Roger Payne, Seiji Ohsumi, Sam Ridgway, and Bernd Würsig. We would be remiss if we did not mention Dave and Melba Caldwell, Hideo Omura, Masaharu Nishiwaki, Graham Ross, Bill Watkins, Bill Schevill, Victor Scheffer, Steve Leatherwood, Bill Evans, "Woody" Wood, and Ken Norris.

We are enormously grateful to Alex Werth for his odontocete insights, and even more so for his willingness to discuss them with us. Like sperm whales, beaked whales are suction-feeders extraordinaire (not to mention *exclusively*).

We benefitted greatly from comments on the book manuscript from Robin Baird, Bill Perrin, Bob Pitman, Anton van Helden, Alan Baker, and

Ken Balcomb. Colin MacLeod has spent the past fifteen years or so collecting information on the lives of beaked whales, and in that time he has produced an impressive body of work almost exclusively in peer-reviewed scientific journals. Colin allowed us to incorporate many of his interpretations (such as males raking each other with their teeth as one of them swims upside-down) into this book. Without Colin's original thinking and enthusiastic support, this would have been a weaker book—and not nearly as much fun to write.

We are very grateful to Vincent Burke, our patient editor at Johns Hopkins University Press. Vince suggested that we work together on this book, joining, as he put it, "the charisma of the artist-naturalist with the detailed scholarship of the Smithsonian scientist."

While very few photographs have been used in this book, we did learn much from looking at those we reviewed. For sending in photographs we thank the Hal Whitehead Lab at Dalhousie University and specifically Hilary Moors-Murphy, CCEMA (Coordinadora para o Estudo dos Mamíferos Mariños), Lisa Denning of Ocean Eyes Photography in Hawaii, John Ford, Ari Friedlander, Noriko Funasaka, Pete Gill, Eliza Muirhead, Keith Mullin, Todd Pusser, Karen Stockin, Jean-Pierre Sylvestre, Wanganui Museum, and Chloé Yzoard.

We are also grateful to Charlie Whitney Potter, Bob Brownell, Phil Clapham, Bill Walker, Bill McLellan, Toshio Kasuya, John Heyning, Merel Dalebout, Ewan Fordyce, and Ed Mitchell.

We dedicate this book to our wives, Stephanie (Richard) and Becky (Jim).

GUIDE TO THE BEAKED WHALES

The species that follow are arranged alphabetically by scientific species name. Included for each species is a painting of an adult male and an illustration of the mandible of an adult male. The mandible drawings give the position and shape of the teeth, which are key characters in differentiating the species.

Arnoux's Beaked Whale

Berardius arnuxii Duvernoy, 1851

If you were to look at a drawing of this species, you might be inclined to say, "It looks like a bottlenose *dolphin*." It has a prominent beak and a bulging forehead, and it is darker above and lighter below. But then somebody would tell you that this animal is 32 feet long, *four times* the length of the average bottlenose dolphin.

Arnoux's beaked whale is the only four-toothed whale found in the waters of the Southern Hemisphere. It has been recorded from Australia, New Zealand, Argentina, Brazil, Chile, the Falkland Islands, Tierra del Fuego, the South Shetlands, South Africa, and the Antarctic continent. Unlike most of the other beaked whales, both males and females of this genus have erupted teeth. There are two pairs, roughly triangular in shape and laterally compressed; the anterior pair is considerably larger than the posterior pair. The lower jaw extends beyond the upper, and often the forward pair of teeth is visible when the animal's mouth is closed. (Because there are no opposing teeth, barnacles sometimes adhere to these teeth.) According to McCann (1975), *B. arnuxii* may obtain a length of 33 feet (9.9 m). The Smithsonian has a record from an animal in the British Museum of 31 feet 7 inches (9.63 m). The whale is dark above and lighter below, often conspicuously marked with white scars and patches. McCann's 1975 photographs of an old male stranded at Pukerua Bay, New Zealand, show an animal literally covered with spots, scars, scratches, and welts. McCann attributed these to the "battle-teeth" of rival adult males, but he also said that they might have been inflicted by killer

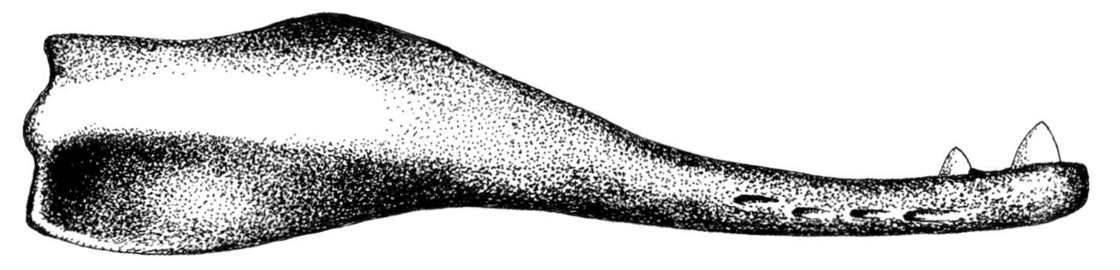

Mandible of *Berardius arnuxii*.

whales, lampreys, or whale suckers, or perhaps they are the "'jab' wounds caused by the tip of an opponent's rostrum."

The name of this species is derived directly from the names of the people involved in its discovery. Captain Berard commanded the French corvette *Rhin,* on which the type specimen was transported to France (described in detail by Duvernoy in 1851), and Arnoux was the surgeon on board who provided a brief description of the animal, which had been collected in New Zealand. For reasons long lost to history, the "o" in Arnoux's name was omitted in the original description, and despite the appearance of various spellings of the name (for example, *arnouxi* or *arnouxii*), the rules of zoological nomenclature require that the original spelling be retained, even though a spelling error was committed. (In taxonomy this kind of error is known as a *lapsus calami,* or a "slip of the pen.") According to Rice's 1998 *Marine Mammals of the World,* the correct name of this animal is *Berardius arnuxii.* To complicate our limited knowledge of this rare whale further, practically everyone who discusses it uses a different common name. It has been variously referred to as the "southern four-toothed whale" (Moore 1968); the "southern beaked whale" (Nishiwaki 1972); "Arnoux's beaked whale" (Brownell 1974); and the "southern porpoise whale" (McCann 1975). A more convincing argument for the use of scientific nomenclature would be hard to find.

At sea this species can be confused with the southern bottlenose whale (*Hyperoodon planifrons*), and immature specimens could be mistaken for almost any species of beaked whale known to inhabit the Southern Ocean. The forehead of the male is not as bulbous as that of the southern bottlenose whale, and the former is usually a larger animal with a smaller dorsal fin.

As this species is known from fewer than fifty specimens, almost all of them strandings or skeletal remains, there is hardly any information on the biology of the living animal. An early report of a stranded animal (Haast 1870) noted a large number of squid beaks in the stomach. In one of the few descriptions of a living specimen, Taylor (1957) saw a single specimen trapped in the closing ice in Graham Land, Antarctica, along with 60 killer whales and about 120 minke whales. "The author only saw the single specimen of *Berardius* on 14.8.55. It was jumping, and four jumps were recorded in 1–5 minutes . . . The Argentines recorded *Berardius* in early September and after taking photographs, they fired repeatedly and presumably killed it."

Recent discoveries by Japanese cetologists indicate that the gestation period for *Berardius bairdii,* the close northern relative of *B. arnuxii,* is perhaps as long as 17 months (Kasuya 1977). Because they are so similar in other respects, the two species of this genus probably share this extraordinary distinction: they would appear to have the longest gestation period of any cetacean. Newborn calves of *B. bairdii* are about 15 feet (4.6 m) in length, and the newborn *B. arnuxii* is probably about the same size.

Baird's Beaked Whale

Berardius bairdii Stejneger, 1883

In 1883 Dr. Leonhard Stejneger described a new beaked whale species from a skull found in the Commander Islands, the westernmost extension of the Aleutian chain. He called it *Berardius bairdii* because he recognized that it was similar to *Berardius arnuxii*, the Southern Hemisphere version, which had been described in 1851. The two *Berardius* species are in fact very similar (*B. arnuxii* may be a little smaller); it is only their discontinuous, semitropical distribution that separates them. *B. bairdii* is found only in the North Pacific Ocean from about 23°N latitude into the Bering Sea, while records of *B. arnuxii* are only from about 34°S latitude to the Antarctic ice pack and continent (Balcomb 1989).

The species was named for Spencer Fullerton Baird, an American naturalist and second secretary of the Smithsonian Institution (US National Museum). In Japanese the species is known as *Tsuchi-kujira*. According to Omura and colleagues (1955), a *tsuchi* is a wooden hammer whose shape closely resembles a bottle, and *kujira* means "whale." It is this name, when translated into English as "bottlenose whale," that seems to have led to mistaken reports of *Hyperoodon* in the North Pacific. (In 1971, Nishiwaki and Oguro wrote that "the whales caught in Abashiri area [north of the island of Hokkaido] . . . should be *Hyperoodon ampullatus*." They also noted that "further study is required.") Recently, another form of *Berardius* has been discovered in the North Pacific and may turn out to be specifically distinct from *B. bairdii* (Kitamura et al. 2013).

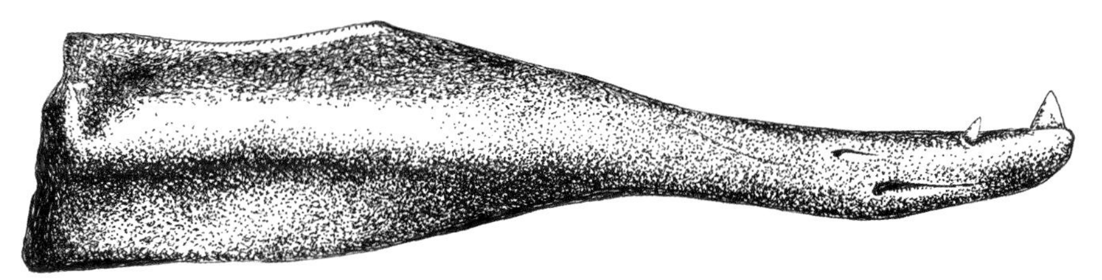

Mandible of *Berardius bairdii*.

Baird's beaked whale is the largest of all the beaked whales, reaching a maximum length of 42 feet (12.8 meters), recorded from a specimen that stranded at Centerville, California, in 1904 (True 1910). On average, the females are larger; males reach a maximum length of about 39 feet (Ralls 1976). Tomilin (1957) recorded a 35-foot specimen that weighed 7.5 tons. Like its southern relative *B. arnuxii*, the adult males and females have four teeth in the lower jaw and none in the upper. The anterior pair is about twice the size of the posterior pair and project obliquely forward well beyond the upper lip. The smaller posterior teeth are concealed when the animal's mouth is closed. As the narrow beak widens somewhat at the end, Tomilin compared it to a duck's beak. The head is small, accounting for only one-eighth of the total body length, and the forehead bulges, but not as much as it does in *Hyperoodon*. The flippers are small and rounded, but the flukes, which usually do not have a median notch, are large, equal to perhaps one-fourth of the body length of the animal. The color has been described as black, dark gray, slate gray, or brownish gray, and there are usually references to lighter undersides and some patches of lighter color, either on the forehead or on the dorsal fin. Another diagnostic feature is the presence of numerous scars and scratches covering the back and sides.

Full-term fetuses have been measured at 13.7 feet (4.20 m), and it has been estimated that this animal is 15 feet (4.50 m) long at birth (Mead 1984). Earlier estimates of the gestation period (for example, Omura et al. 1955; Tomilin 1957) gave a figure of 10 or 11 months, but in a 1977 study, Kasuya concluded that the actual gestation period is 519 days, or 17 months. (This is the longest known gestation period for any cetacean, and of all other mammals, only the elephants, with gestation periods of over 600 days, surpass it.) Through the same examinations of the dentinal layers of the teeth, Kasuya also estimated the life span at about 70 years.

These whales congregate in offshore waters at least 3,300 feet (1,000 m) deep and rarely enter shallower waters. They are often seen in groups of 10 to 30 animals, comprising males, females, and juveniles. (A school of 18 individuals was photographed off the coast of Japan [Nishiwaki and Oguro 1971], one of the few instances in which living beaked whales have been photographed and identified en masse.) Although they are deep-water inhabitants, they do occasionally strand, and beached animals have been found in the Aleutians, British Columbia, Washington, California, and Japan. In July 2003, the shark-bitten carcass of a large beaked whale was found decomposing on the shore of Ocean Beach, San Francisco. Scientists from the California Academy of Sciences (CAS) identified it as *B. bairdii*, and CAS specimen collector Ray Bandar removed the skull for the museum's collection. After one that had been collected in 1923 at Santa Cruz, this was the second Baird's beaked whale skull in the Academy's vertebrate collection.

Like most beaked whales, *Berardius bairdii* is a *teuthophage*, a squid eater. Its main prey species are *Gonatus fabricii* and *Onychoteuthis* in California waters (Rice 1963), but there are records of it feeding elsewhere in its range on deep-sea fishes, mackerel, sardines, pollock, and sauries (Nishiwaki and Oguro 1971).

From 1985 to 1991, researchers William Walker, James Mead, and Robert Brownell analyzed the stomach contents of 107 Baird's beaked whales taken by Japanese whalers in the coastal waters of Japan and another 20 killed in the southern Sea of Okhotsk, north of the island of Hokkaido. Of their 2002 report, they wrote that "It not only represents the first detailed quantitative prey species description for this species of whale, but also constitutes the most comprehensive diet study conducted on any member of the family Ziphiidae." The diet of the whales taken off the Boso Peninsula, in the vicinity of the whaling port of Wadaura, consisted primarily of benthopelagic fishes, especially codlings and grenadiers, while the food of the Sea of Okhotsk whales was 87% cephalopods, of varying sizes and species. In both regions, the water gets to be more than 3,000 feet deep, ideal conditions for the deep-diving *Berardius*.

Its surface behavior has been compared to that of the sperm whale: it takes a number of short breaths before descending for deep hunting forays. There are reports of Baird's beaked whales staying submerged for over two hours, which would place it among the record breath-holders for all cetaceans. (Sperm whales can remain below for an hour and a half.) One account mentions an individual that was harpooned and took 900 meters (2,952 ft) of line straight down. We have been told that the Japanese who hunt these whales regard them as the most dangerous of all whales to harpoon, because of the incredible speed with which they can descend. Pike (1953) observed a male struck by a harpoon that "dived straight down at amazing speed, taking with it 500 fathoms of line." When the animals surface again, they first show their foreheads and then breathe and roll, with the small dorsal fin rarely breaking the surface. Tomilin told of Soviet whalers who observed these whales "breaching" in the waters of Kamchatka, where they "jumped obliquely displaying three-fourths of the body in the air." Males are usually seen covered with myriad scratch marks, as are most of the beaked whales, but until someone actually observes the animals engaging in combat or wrestling with squid that can scratch their skin in parallel lines, we can only speculate about the cause of these multiple surface wounds.

In 1931, R. C. Andrews wrote, "When I was in Japan in 1910, I saw a photograph of a whale which was said to occur at certain times of the year only in Tokyo Bay, and when a skeleton was finally secured . . . the whale was found to represent an extremely rare species, *Berardius bairdii* which had been taken only in Alaskan waters" (p. 266). In fact, the animal has been hunted for centuries by the Japanese, but catches have never been particularly high because of the rarity of the species. From 1969 to 1977 the Japanese fishery, mostly off the Boso Peninsula on the Pacific side of the main island of Honshu, took some 622 animals, or an average of 69 per year. (During this same period, the International Whaling Statistics reported no whales of this species taken by any other nations.) In the recent past, the Soviets hunted them in the western North Pacific off Kamchatka and the Sea of Okhotsk, but Tomilin reported in 1957 that they found it uneconomical and "taken in the waters of the U.S.S.R. only by chance." He also claimed that the fat and flesh are inedible, but "boiled meat can be safely fed to fur animals." Tomilin (1967: 426) quoted Krasheninnikov (1755), who wrote that the fat does not stay in the stomach but "flows out imperceptibly through the inferior passage." Unfortunately, Tomilin (1967) did not include a bibliography listing the details of any of the papers that he cited.

Japanese whalers killed some four thousand *Tsuchi-kujira* before the 1986 moratorium on commercial whaling, and they are still killing them, selling the meat for human consumption. In October 2013, the New Zealand newspaper *Marlborough Express* published a report by Jared Nicol, entitled "Whale Meat at $22 a Kilo," wherein Nicol had visited the Japanese whaling port of Wadaura and watched the workmen carving up a 33-foot-long *Tsuchi-kujira*. Nicol wrote,

> Sotobo Whaling Company flensers annually carve up to 26 *tsuchi kujira* . . . at an abbatoir in the port town of Wadaura on Japan's Pacific coastline. The meat and blubber is sold on-site or packed with ice and shipped to restaurants and supermarkets around the country . . . The fat is usually used to make an oily fish soup, while the deep red meat—particularly the richest cut which runs down by the whale's tail—is fried with cabbage and tastes like a gamey mix of tuna and beef . . . The big bricks of meat sold that day for about 1700 yen or NZ$22 per kilogram . . . A whole whale can be worth about five million yen or NZ$65,000.

The Japanese argue that they are a “small cetacean species” and not protected under the International Whaling Commission moratorium on commercial whaling, even though they can get considerably larger than minke whales, which *are* included in the moratorium. Japan’s ministry of agriculture, forestry, and fisheries permits a catch of 65 of these whales every year: 52 from the Pacific coast, 4 from the Okhotsk Sea, and 10 from the Sea of Japan. The meat and blubber food products of these and other whales have been found to contain high levels of mercury and other pollutants, such as PCBs.

Northern Bottlenose Whale

Hyperoodon ampullatus (Forster, 1770)

Adult males have a bulging forehead, a small but conspicuous beak, and a mottled brownish or yellowish coloration, which becomes almost white in the oldest and largest specimens. Females are usually smaller and darker in color and do not exhibit the overhanging forehead bulge. The maximum size is 30 feet (9.15 meters) for adult males and 24 feet (7.3 m) for females. Males are also heavier in build; one nineteenth-century observer wrote that "the females, as is proper, have much more graceful outlines" (Southwell 1883). A 21-foot (6.6 m) specimen weighed 4,480 pounds (2200 kg) (Nishiwaki 1972). Although juveniles are dark in color (they have been described by Winn et al. [1970] as "dark chocolate brown"), they begin to lighten as they mature. Most cetaceans darken appreciably in color when they die, so many descriptions of this species from stranded specimens describe the animal as black. According to observations of live animals at sea (D. Gray 1882; Ohlin 1893; Winn et al. 1970), the animals are tan or brownish, often with lighter spots and scratches. Males are usually white or cream-colored on the forehead bulge, and occasionally an all-white animal is seen. Ohlin wrote, "I have seen but once such a 'whitefish.'" In some females there is a light ring around the neck, and Ohlin claimed that the Norwegian whalers call these specimens *ringfiskar*, or "fishes with a ring."

In the males the maxillary crests of the skull develop into an exaggerated pair of bony ridges, a development that does not occur in females. In some illustrations (for example, Gray 1882) the forehead is flattened on the forward surface of the males, whereas the females have a sloping profile more like that

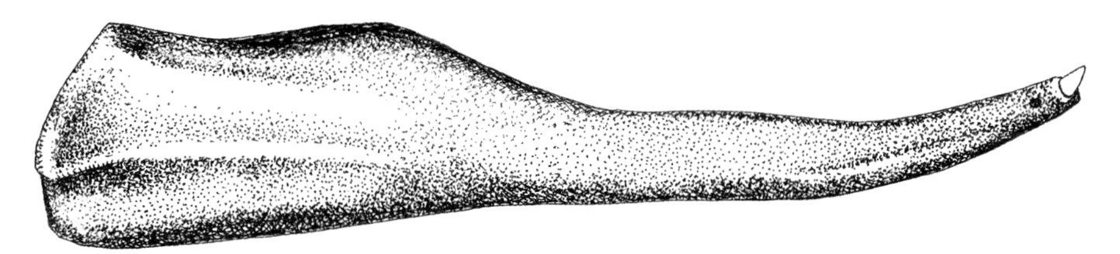

Mandible of *Hyperoodon ampullatus* male.

Lithograph of a 764 cm female northern bottlenose whale that stranded at Zantvoord, Netherlands, on the 24th of July, 1846 (Vrolik 1848, plate I).

of the other beaked whales. This dimorphism is so pronounced that early anatomists believed that the variations represented two distinct species. In an article entitled "On the Genus *Hyperoodon*: The Two British Kinds and Their Food," J. E. Gray (1860) wrote: "The structure and form of the two skulls is so different, that it is much more likely that they should be referable to two very distinct genera than to species of the same genus." It was later demonstrated that the two species, *H. latifrons* and *H. rostratus*, were actually the males and females of the same species (Flower 1882 citing Van Beneden and Gervais 1868–79). The name *Hyperoodon* can be defined as "beyond teeth" or "over teeth," and it refers to the overhanging forehead in the males. *Ampullatus*, which means "flask-shaped," describes the animal's head, characterized by a narrow snout protruding from a large, well-defined fleshy melon.

Males and females of this species have but one pair of teeth, located at the tip of the lower jaw. Elongated and conical in shape, these are set in the jaw pointing forward at an angle. The teeth are not a particularly reliable means of identification, because in some specimens the teeth do not erupt at all, and in other, older animals they might fall out or wear down completely. Some specimens have been described as having toothpick-sized teeth in addition to the apical pair, and some even have another pair of teeth embedded in the gums behind the first pair. In the X-ray examination of a small female that stranded at Waterford, Ireland, in 1938, investigators discovered a complete set of tiny, vestigial teeth buried in the upper and lower gums (Fraser 1953).

The northern bottlenose whale is a North Atlantic species. One report tells of one being taken in the waters off the northern Japanese island of Hokkaido (Nishiwaki and Oguro 1971), but this seems to be a case of mistaken identity. (The even larger beaked whale, *Berardius bairdii*, is commonly fished in these North Pacific waters, and it would be quite possible to confuse them, especially when translating information from Japanese into English.) It is a cold-water, deepwater species, which spends the spring and summer around the edges of the pack ice and moves south in the fall and winter. The Norwegians have hunted this animal off East Greenland and Labrador, usually in waters deeper than 3,300 feet (1,000 m). The species does not usually occur in shallower waters, but it does strand with some frequency on both sides of the North Atlantic. Fraser (1974) recorded some sixty-six strandings of this species from 1913 to 1966, and Duguy (1977), writing of the coasts of France, said that "strandings occur regularly but without being frequent." The Cetacean Distributional Database at the Smithsonian Institution has at present 450 records of the northern bottlenose whale, of which 332 are strandings. There are numerous records of northern bottlenoses stranding in North America, from various Canadian locations to New England, with the southernmost confirmed location being Rhode Island (Mitchell and Kozicki 1975). On January 20, 2006, a strange long cetacean, originally thought to be an oversized dolphin, was spotted swimming upstream in the Thames in central London. Attempts to turn the 18-foot-long immature female bottlenose whale (for that is what is was) back toward the sea all failed, and within two days she was exhibiting signs of panic and stress. Divers entered the water, trying to turn her around, but she thrashed wildly and died. Five years later, the skeleton of the "Whale That Swam up the Thames" was placed on exhibition at the Natural History Museum at Tring, Hertfordshire.

The northern bottlenose congregates in small groups of five to fifteen, but much larger schools have been observed (Nansen, quoted in Benjaminsen and Christensen 1979). The deepwater habitat of this species is probably related to the habits of its predominant prey, the squid *Gonatus fabricii*; captured specimens have been found with over ten thousand squid beaks in their multi-chambered stomachs (Ohlin 1893). Species members have also been known to feed on herring and other kinds of fish, and some bottom feeding is indicated by the existence in their stomachs of such bottom-dwelling creatures as starfish and holothurians (sea cucumbers) and even sand and stones. Ohlin observed "a thin covering of lime upon its beak from its rummagings in the muddy ground of the sea."

The northern bottlenose is reputed to be among the deepest diving of all cetaceans (Scholander 1940). David Gray, an articulate whaleman, observed in 1882:

> They have great endurance, and are very difficult to kill, seldom taking out less than from three to four hundred fathoms of line; and strong full-grown males will run out seven hundred fathoms, remaining under water for the long period of two hours, coming to the surface again as fresh as

> if they had never been away; and if they are relieved of the weight by the lines being hauled in off them before they receive a second harpoon and a well-placed lance or two, it often takes hours to kill them. They never die without a hard struggle, lashing the sea white about them, leaping out of the water, striking the boats with their tails, running against them with their heads and sometimes staving the planks in, frequently towing two heavy whale-boats about after them with great rapidity.

When Ohlin observed them in the "Arctic Sea" in 1891, he saw them dive regularly for one to two hours and then "be seen to appear in the immediate vicinity whence they dived." Other firsthand observations corroborate this achievement, and Benjaminsen and Christensen recorded diving times (of unstressed animals) of fourteen to seventy minutes. Harpooned bottlenoses are said to "dive straight downward at tremendous speed, and have been known to take out five hundred fathoms of line in two minutes" (Andrews 1931: 260). The ability to dive deeply and remain submerged for extended periods of time has been demonstrated by only a few other species of cetaceans, among them the sperm whale and the giant beaked whales of the genus *Berardius*. (Other species, such as the bottlenose dolphins and the pilot whales, have demonstrated deep-diving capabilities under training, but there is no evidence that they descend to extreme depths in the course of their normal feeding.)

Fridtjof Nansen (1861–1930), famed Norwegian explorer and scientist who made the first crossing of the Greenland ice sheet in 1888–89, sailed aboard the sealing vessel *Viking* in the spring of 1882. Their quarry was the "saddleback" (now known as the harp seal), but in *Hunting and Adventure in the Arctic*, Nansen included several tales of bottlenose whaling off the coast of northern Norway, one of which was told to him by a seaman named Markussen:

> One fine day I saw a lot of bottlenoses about, so I rigged up a boat with a harpoon and took three whale-lines just to be on the safe side . . . We soon fell in with a fine fellow who came up right ahead of the boat. When I stuck the harpoon into him, he made a hell of a splash and then sounded; the line ran out so fast that you could smell burning . . . The first line ran out and the second soon followed; then he started on the third line, and it ran out every bit as fast as the other two had done . . . Out went coil after coil, and the pace seemed to be just as fast as ever. And when he had taken the lot of it he pulled the boat under too, straight on without

a stop; down it went and left us kicking about in the water." [They were picked up by their ship, and Markussen continued] . . . "Them fish are the very devil to stay under water. Though the sea was like glass and we kept a sharp look-out from the crow's nest all day in the hope of seeing our boat, neither boat nor fish did we ever see again. He certainly didn't come up anywhere between us and the horizon.

Hal Whitehead of Dalhousie University in Halifax, Nova Scotia, probably the world's foremost authority on sperm whales, has also studied the population of *Hyperoodon ampullatus* in "the Gully," an underwater canyon 220 miles east of Halifax. Whitehead and his colleagues have been counting and photographing the northern bottlenoses of the Gully since 1988, and in a 1997 review of these observations, they estimated the population at approximately 230 animals. Also working in the Gully, Sascha Hooker and Robin Baird (then also of Dalhousie University) obtained the first dive profiles of any beaked whales. With a crossbow, they affixed suction-cup time-depth recorders to the whales, and recorded the results. From a total of 56 sonar recordings, they hypothesized that *H. ampullatus* "may make greater use of the deep portions of the water column than any other mammals so far studied" (Whitehead 2003). The deepest dive was recorded at 1,453 meters—4,767 feet down. These dives, some of which lasted for 70 minutes, took them to where the 2-foot-long squid *Gonatus fabricii* (their favorite food item) could be echolocated and sucked up (Hooker and Baird 1999a).

Even though this species is well known—probably *best* known—from the waters of Nova Scotia, an indication of its rarity is this story, published

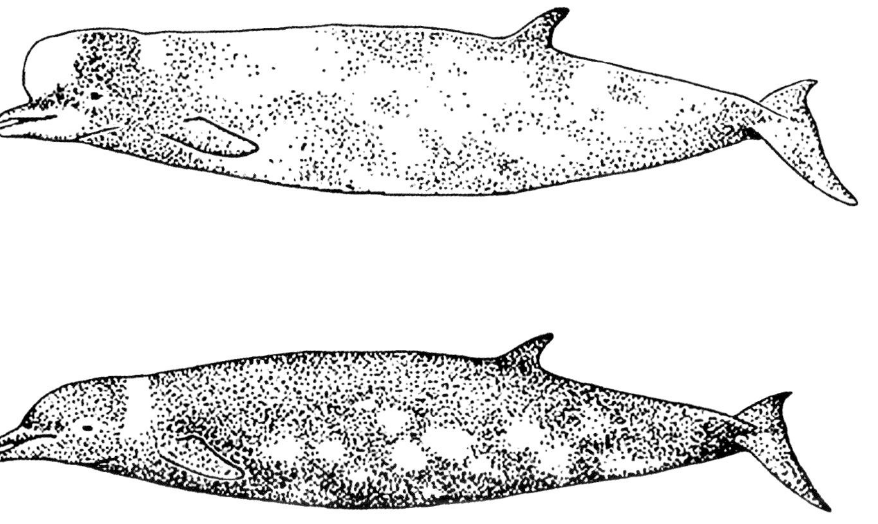

Male and female northern bottlenose whales look so different that for many years scientists believed they were two different species. Males are lighter in color and have a bulbous forehead.

by the Canadian news service (i.e., CBC News) on October 11, 2013: "Rare Bottlenose Whale Pops Up Again on Eastern Shore, N.S." An adult, estimated at a length of 9 meters (30 ft), was spotted in Spry Bay, on the eastern shore, far from its normal offshore, deep-water habitat. It was seen swimming at the surface and occasionally breaching, making for an unquestionably accurate identification. According to Hal Whitehead, "The chances of seeing a northern bottlenose whale from the coast anywhere in the world are virtually zero." He suggested that the whale was so close to shore because it was probably sick or injured, and he was correct. Rescuers tried to turn the whale toward the open sea, but it refused to go and within three days was reported dead in Spry Bay.

The nose of the sperm whale—with its valves, sacs, tubes, and oil reservoir—is unique among cetaceans, but there are some similarities between *Physeter* and *Hyperoodon* that are probably related to their deep-diving propensities. Whereas the sperm whale has the spermaceti organ, most fully developed in the large bulls, *Hyperoodon* males develop a bulging forehead, exaggerated crests on the skull, and, between these crests, "a solid lump of fat similar in shape to, and twice the size of, a water-melon" (D. Gray 1882). Ohlin (1893) described this organ as being composed of "tissue resembling a bee-hive. The rooms between the bands are filled with a clear, thin-floating oil." Female bottlenoses, which lack the bony crests and the bulging forehead, have only a small amount of thin, yellow oil in their heads. The oil from the bottlenose appears to be "of fine quality and hardly distinguishable (if at all) from sperm" (Thompson 1919). Norwegian whalers sometimes obtained as much as 440 pounds (200 kg) of spermaceti and over 2 tons of oil from a large bottlenose (Tomilin 1967). The same author also noted that "the meat is unfit for human consumption. It must be boiled to eliminate its laxative properties, and can then be used as food for dogs and fur animals at state breeding farms." The spermaceti organ of the sperm whale is somehow involved with deep diving (see M. R. Clarke 1978a,b,c for elaboration), and it is probably so with the northern bottlenose as well.

Sound production in the sperm whale is an activity still shrouded in mystery, but most cetologists agree that the clicks, wheezes, and knocks are produced in the nose, wherein lies the massive spermaceti organ. In the bottlenose the maxillary crests of the skull almost certainly serve an acoustic function. Norris (1964) wrote that "the scoop-shaped bones of the forehead . . . look for all the world like parabolic surfaces whose focal points lie in the general area of the soft anatomy of the forehead and hence might act as sound reflecting and focusing devices." Underwater sounds have been recorded from

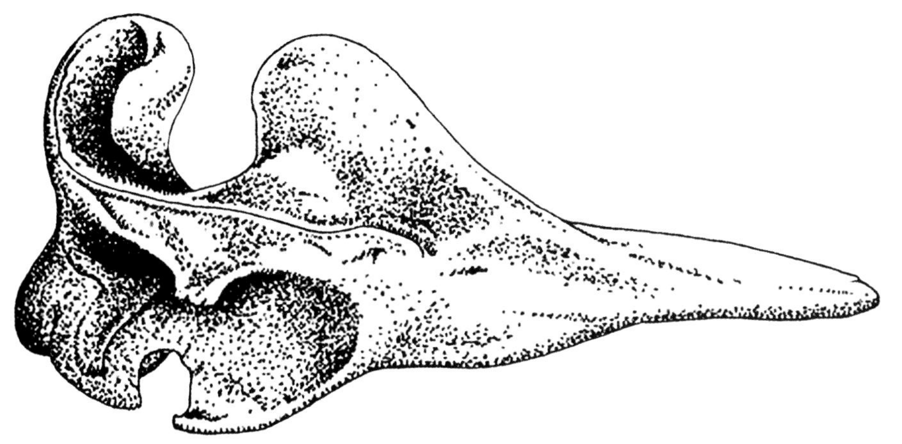

Lateral view of the skull of a male of *Hyperoodon ampullatus*, showing the exaggerated development of the maxillary crests.

this species by Winn and colleagues (1971)—the first recorded sounds ever described for a species of beaked whale—and they consist of "discrete frequency whistles, sweep frequency chirps, and possible burst-pulsed (modulated) tones." The surface sounds of this animal are also noteworthy. Tomilin (1967) claimed that "the bottlenose can be unmistakably distinguished from other cetaceans by the sharp noise of its breathing: this noise is brief [and] of a brassy pitch."

Whitehead, in a book on sperm whales that was published in 2003, gives a current synopsis of the behavior and vocalization of the bottlenose whale. At birth the bottlenose whale is reported to be 10–11 feet (3.0–3.3 m) in length; and although the gestation period is unknown, Benjaminsen (1972) estimated it at twelve months. The species is known for its strong schooling instinct, and females have been observed protecting their calves from boats or "standing by" an injured or distressed adult. (The Norwegian whalers used this inclination to their advantage: since the school would not leave a wounded comrade, they often shot the whole school at once without having to chase them.) The northern bottlenose has also been described as playful and inquisitive, approaching boats from great distances, jumping, and tail-slapping within 25 feet of the vessel (Winn et al. 1971). Like most of the beaked whales, this species is often found with scratches and scars, and as usual, their specific origin is unknown. Benjaminsen and Christensen (1979) suggested that the forehead scars "may have been caused by the ice when the whales pushed through for breathing, although other explanations are possible." Because this species has what McCann (1974) called "battle-teeth"—that is, teeth located at the extremity of the jaws—he felt that the scars are caused by fighting males. Nishiwaki (1972) suggested that "the large forehead is probably used in fighting for the females," and there is also the possibility that some of the scars are caused by the suckers of squid. Gowans and Rendell (1999) observed that because the "battle teeth" of the northern bottlenose whale are not as highly modified as those of other beaked whales, the males often resort to "head-butting," using the large, flattened melon as a weapon. The same might well hold true for the very similar *Berardius*.

Ohlin (1893) believed this species to be the largest of the odontocetes next to the sperm whale, but it is not, for *Berardius* grows substantially larger. However, it has a long and productive history in the small-whale fishery. Captain David Gray of Peterhead, Scotland, was the first to seek these whales in the waters north of his homeland. In 1881 he brought in the first cargo of bottlenose oil to Dundee, and the following year he killed 200 bottlenoses. The Norwegians soon joined in this fishery, which was, after all, off their own coast; by 1891, 70 Norwegian ships had taken 3,000 animals. Of this early period of bottlenose whaling, Mitchell (1977) has written, "In the period 1886–1911, catch efficiency undoubtedly increased as a rapidly evolving whale-catching technology saw the development of: whaling cannons, experimentation with chemical, explosive and other harpoon heads, bombs and grenades, lighter and stronger foregoers, spring systems to prevent breakage of lines when playing whales, motorised vessels, and many other innovations."

The bottlenose was hunted from small schooners of 30 to 50 tons and was shot directly with guns mounted on these vessels. (Unlike the catcher boats used for the larger species—which had a single harpoon cannon on the bow—these boats had guns all around, to take advantage of the bottlenose's habit of approaching boats from all points of the compass.) According to Leatherwood and Reeves (1983), "Norwegian sealers killed more than a thousand in the Greenland Sea in 1885, and an average of 2,500 per year during the 1890s. Although the fishery ceased in the 1920s, it resumed after the Second World War and continued into the later 1960s . . . Oil and animal food were the main products." By 1920, when the first phase of the fishery began to decline, approximately fifty thousand bottlenose whales had been killed (Jonsgard 1955). Mitchell and Kozicki (1975) reported a Canadian fishery out of Nova Scotia that took 87 bottlenoses from 1962 to 1967, and the species was also hunted by the Danes and the Icelanders.

Nansen (1925) contributed to the controversy about whether the meat of beaked whales is edible when he wrote:

> The flesh of this whale is quite palatable, like the flesh of other species of whales, but its blubber and train-oil, being strongly aperient [having

laxative qualities] cannot be eaten. This fact was known to the Norwegians in very early times and is mentioned in *The King's Mirror* (about 1240 A.D.), where we read the description of the whales around Iceland: "Furthermore there are two kinds of whale, one is called "*And-hval*" (Duck-whale) and the other the "*Svin-hval*" (Pig-whale) . . . these fishes are not edible, for the fat which runs off them cannot be digested by human beings or any other animal, for it runs right through them and likewise through wood; indeed, it is well nigh impossible to preserve it if it stands for any length of time, even when stored in horns."

Southern Bottlenose Whale

Hyperoodon planifrons Flower, 1882

This is the Southern Ocean version of the bottlenose whale, and it resembles its northern relative in most particulars. Mature males have a bulging forehead that overhangs the base of the beak, and there are two teeth at the tip of the lower jaw. Both males and females have teeth, but those of the females are usually buried in the gums. There does not seem to be the same sequence of color change from dark to light during the animals' maturation as there is in *H. ampullatus,* since most descriptions refer only to their dark color. Gaskin (1968) described it as "generally dark brownish-black dorsally and greyish white ventrally. White markings may be present on the back and flanks." The southern bottlenose whale reaches a maximum length of 25 feet (7.6 m); this is somewhat smaller than the northern species. Unlike its northern relative, for which information has been gathered from a prospering fishery in the North Atlantic, the southern variety is known primarily from the examination of stranded specimens. The Latin name *planifrons* means "flattened brow," referring to the shape of the head in the living animal or, perhaps, the shape of the maxillary crests of the skull. In Spanish it is known as *ballena pico de botella,* or "bottle-beak whale."

This species is found only in the Southern Ocean, from the Antarctic pack ice to southern Brazil (Gianuca and Castello 1976) on the east coast of South America and to Valparaiso, Chile, on the west. It has been found off Tierra del Fuego (Goodall 1978), Uruguay, the southern coasts of Australia, New Zealand, the Falkland Islands, Heard Island, Namibia, New Zealand, and South Africa. In the waters of the southern latitudes and the Antarctic

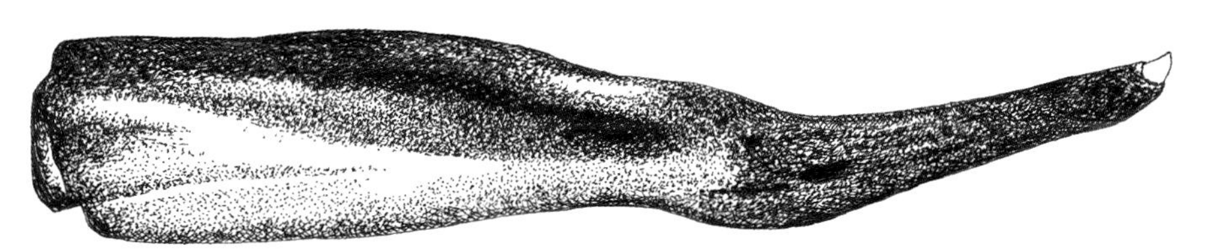

Mandible of *Hyperoodon planifrons.*

this species could be confused with the giant bottlenose, and there does not seem to be an accurate way to differentiate them at sea. A skull examination would show that the bottlenose has greatly exaggerated crests—particularly in mature animals—whereas in the giant bottlenose, the crests are less developed and the skull is more symmetrical. Also diagnostic would be the number of teeth in the end of the jaw: the bottlenose has only two teeth and the giant bottlenose, four. These teeth were described by McCann (1961) as "large, obovate, or oboconical bodies surmounted by mammilate enamelled apices"—in other words, egg shaped with enameled tips. There appears to be some confusion regarding the actual appearance of the teeth in the bottlenose, as Norman and Fraser (1938) described them as "conical but expanding somewhat at about half their length then narrowing slightly to the lower end." The illustrations of the teeth of the northern bottlenose in Moore (1968) correspond to this description, but not at all to McCann's "oboconical bodies." Moore has written that the teeth of males and females are quite different, and "the evidence is strong that the southern hemisphere species of *Hyperoodon* has achieved striking sexual dimorphism in the shape of the teeth." In addition, there is a marked change in the shape of the teeth as the animals mature.

Since we know so little about the habits of this animal, we can assume only that its habits are similar to those of the northern bottlenose. It is probably a squid eater; it is capable of deep dives; and its forehead, or "melon," contains a reservoir of waxy spermaceti, which Baker (1983) says "may act as an acoustic lens for directional beaming of echolocation signals, and prey stunning sound bursts." A full-term fetus measured 9 feet (2.7 m), but nothing is known of breeding or parturition behavior. Hale (1939) reported that a specimen stranded alive in Victoria, Australia, "made a grunting noise like a pig;" and Lionville, who observed the animals on the French Antarctic Expedition of 1908–10, said it sounded like a trumpet when it breathed.

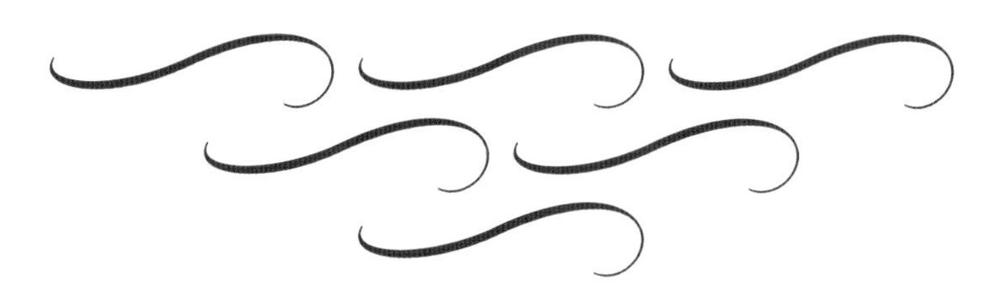

Longman's Beaked Whale

Indopacetus pacificus (Longman, 1926)

In his 1980 *Book of Whales,* Ellis wrote: "This may be the least-known large animal in the world today. It is known only from two skulls, the first of which was found on the beach at Mackay, Queensland (Australia), in 1882, but not described until 1926. The second skull was found in Somalia in 1955, "in a fertilizer factory not far from Mogadishu: it had been collected a few weeks before by local fishermen at Danane" (Azzarolli 1968). The Queensland skull is almost 4 feet in length, so it was assumed that this was one of the larger of the beaked whales. At first it was thought to be a rare member of the *Mesoplodon* genus, but in 1972, Joseph Curtis Moore assigned it to its own genus, *Indopacetus,* or "Indo-Pacific whale." He wrote, "As *Indopacetus* is known from but two skulls, the character of its external appearance remains to be discovered." Because we knew it existed, it seemed only a matter of time before another *Indopacetus* specimen showed up on a beach somewhere.

Instead, a little cetological detective work showed that *Indopacetus* has been visible for some time—we just didn't know what it was. For at least thirty years, an unidentified species of beaked whale had been reported in the tropical Indian and Pacific oceans. In his 1971 *Field Guide of Whales and Dolphins,* W. F. J. Mörzer Bruyns wrote (regarding *Indopacetus*) of "the possibility that the author observed these animals in the Gulf of Aden and the Sokotra area, being very large beaked whales and certainly not *Ziphius*." Mörzer Bruyns also wrote that "Mr. K.C. Balcomb of Pacific Beach, Washington USA took a photo of a school of 25 beaked whales on the equator at

Mandible of *Indopacetus pacificus.*

165° West [in the vicinity of the Gilbert Islands], which were almost certainly this species." In Balcomb's and other photographs, the animals looked more like bottlenose whales (*Hyperoodon*) than beaked whales, but the two species of bottlenose whales (*H. ampullatus* and *H. planifrons*) are from high northern and southern latitudes, respectively, and had no business being in tropical waters. Both species have bulging foreheads, and although they are born dark gray or brown, they become lighter with age, especially around the head. Males differ so markedly from females (they are considerably larger and have a much more pronounced forehead bulge, commonly known as the "melon"), that early cetologists classified them as two different species. Maximum length for a male bottlenose is 30 feet, and for a female, 23 feet. The mysterious tropical whale was similar in appearance. *Indopacetus pacificus*—the only species in the genus—was about to surface.

In 1999 Pitman et al. published an article in *Marine Mammal Science* with the intriguing title "Sightings and possible identity of a bottlenose whale in the tropical Indopacific: *Indopacetus pacificus*?" (the question mark is part of the title). The article incorporated a collection of photographs (one of which was Balcomb's 1966 picture), and when the photographs and eyewitness descriptions were compared, it was clear that the whale was a bottlenose whale, and it was neither the northern nor the southern version. The photographs clearly show a bottlenose whale whose body color "has been variously described as tan, light brown, acorn-brown, gray-brown or just gray." When an adult female beaked whale stranded in late 1999 in the Maldives in the northern Indian Ocean, it was identified as *Indopacetus*, and comparison with the museum specimens showed once and for all that *Indopacetus* was actually the tropical bottlenose whale. In 2002 in *Natural History*, Robert Pitman wrote: "It had to be out there somewhere. It wasn't just [a] set of car keys that had gone missing—somehow an entire species of whale had been lost for a century." It was not a wraith or a distant sighting at sea; it was a beaked whale whose existence was documented by two skulls that were obviously from the same species, but were different enough from the other beaked whales to warrant their own species designation. Pitman wrote, "For a hundred years, cetologists had nothing to work with but two skulls on the shelf. We now have specimen material from six individuals (including five skulls and one complete skeleton) records of more than two dozen sightings, numerous photographs of large animals in the field, recordings of their vocalizations, and (welcome to the twenty-first century) eight minutes of digital video footage."

And then, in July 2003, a complete redescription of *Indopacetus pacificus* (Dalebout et al. 2003) appeared in *Marine Mammal Science*. It was written by nine authors from New Zealand, Australia, the Maldives, South Africa, Kenya, and California (Pitman was the California contributor). They examined the two original skulls from Queensland and Somalia and added four new specimens to the list: a skull that was found in the National Museum of Kenya; several ribs and vertebrae that had been found on a Natal beach in 1976; a skull, mandible, teeth, ribs, and ear bones of a specimen (mistakenly identified as *H. planifrons*) in the Port Elizabeth (South Africa) Museum; and the adult female that was collected in the Maldives in January 2000. DNA sequencing showed that all these specimens belonged to the same species, *Indopacetus pacificus*, now known as Longman's beaked whale. The range of this species is now known to include "the western reaches of the tropical Pacific Ocean . . . and the western, northern, and southern latitudes of the tropical Indian Ocean." The authors concluded: "The discovery of these four new specimens has extended the known range of Longman's beaked whale and led to the description of its external appearance for the first time. As this species is now known from six specimens, the title of the world's rarest whale must pass to the spade-toothed whale [another beaked whale] *Mesoplodon traversii* (= *M. bahamondi*) which to date is known from only three specimens."

In a comprehensive review published in 2006, Anderson, Clark, Madsen, Johnson, Kiszka, and Breysse added "seventeen new sightings" and reviewed the previously published records, summarizing the available information on Longman's beaked whale in the western Indian Ocean. They noted that "the bulbous melon and distinct beak distinguish this species from other known beaked whales in the tropical western Indian Ocean," and although the color pattern has been well described by Pitman et al. (1999), "the apparent dorsal coloration of live animals varies according to weather conditions. In good sunlight, most animals appear a rich coffee-brown color. Under grey skies, most appear dull grey brown. Secondly, the dark band behind the blowhole is highly variable and sometime indistinct. Thirdly, the lower jaw is normally pale brown or grey, but it is sometime pink." The dorsal fins appear larger and taller than in most other *Mesoplodon* species, and the shallow arch of its back before a long dive distinguishes it readily from *Ziphius*, another brownish beaked whale also found in these waters. Group size in the western Indian Ocean varies from four to eight individuals, somewhat smaller than the groups reported from the eastern tropical Pacific by Pitman et al. The authors wrote that *Indopacetus pacificus* "appears to be more common in the western Indian Ocean, and around the Maldives in particular," but the survey methods were different enough that such a statement may not be definitive.

From the NOAA research vessel *McArthur II* in Hawaiian waters in August 2010, researchers spotted a large group of beaked whales, which were identified as Longman's beaked whales. In their summary of the events, Shannon Rankin, Simone Baumann-Pickering, Tina Yack, and Jay Barlow wrote, "Group size was estimated to be 88 individuals including ~3–5 calves. The group showed a relatively high level of surface activity, and although they did exhibit evasive behavior relative to the ship, including splitting into subgroups, this behavior was less dramatic than is typically observed with this species." A hydrophone array, towed about 300 meters behind the ship at a depth of about 8 to 11 meters at a speed of 10 knots, obtained the first ever sound recording of this species. As the authors described the recording: "Sounds included echolocation clicks and burst pulses. Echolocation clicks were grouped into three categories, a 15 kHz click (n=106), a 25 kHz click (n=136) and a 25 kHz pulse with a frequency-modulated upsweep . . . Burst pulses were long (0.5 s) click trains with approximately 240 clicks." From two unidentified skulls in museum cabinets to 88 individuals swimming and clicking in Hawaiian waters, *Indopacetus* has made a long trip from obscurity to the forefront of beaked whale prominence. Unhappily, a recent incident demonstrates that its newfound celebrity is not necessarily good news for the species.

On March 22, 2010, a juvenile male beaked whale was seen at Hamoa Beach on the Hawaiian island of Maui, and after several minutes of thrashing in the shallows, it died. DNA analysis showed it to be *Indopacetus*. The bad news? The dead whale was found to have been infected with a morbillivirus, a known killer of cetaceans and pinnipeds. In a 1995 summary of the morbillivirus infections of marine mammals, Osterhaus et al. wrote:

> Several disease outbreaks, which have caused the deaths of many thousands of seals and dolphins during the last decade, have now been attributed to infections with newly identified morbilliviruses. Outbreaks in the late eighties amongst harbour seals (*Phoca vitulina*) and grey seals (*Halichoerus grypus*) in northwestern Europe and amongst Baikal seals (*Phoca sibirica*) in Siberia were caused by the newly discovered phocine distemper virus and by a strain of canine distemper virus, respectively. Although closely related these two viruses were not identical. They were more distantly related to the viruses which caused mass mortality

amongst striped dolphins (*Stenella coeruleoalba*) in the Mediteranean sea in the early nineties. This dolphin morbillivirus was shown to be closely related to a virus that was found in harbour porpoises (*Phocoena phocoena*) which had stranded at the coasts of northwestern Europe in the late eighties: porpoise morbillivirus.

Sowerby's Beaked Whale

Mesoplodon bidens (Sowerby, 1804)

In his 1804 *British Miscellany . . . of New, Rare, or Little Known Animal Subjects,* James Sowerby (1757–1822) published the first description of what he called *Physeter bidens*:

> Mr. Brodie (who assisted me with the sketch and description of the rest of the animal) observes that the cuticle on every part of the head and body was perfectly pellucid and satiny, reflecting the sun to a great distance. Immediately under the cuticle, the sides were completely covered with white vermicular streaks, in every direction, which at a little distance appeared like irregular cuts with a small sharp instrument. It was a male animal.
>
> We know of no whale, with only two teeth in the lower jaw, described by any author . . . We cannot be mistaken as to the position of the head in our figure, for the spiracle was sufficiently conspicuous when it was received. We might have called it *Physeter rostratus,* with some propriety; but this might have created confusion. It is however a curious circumstance, that such an appellation would suit better if it were described with the wrong side upwards, which will be easily observed, if the plate would be reversed: and the jaws, in this case, very aptly resemble a bird's beak.
>
> Animal oblong, black above, nearly white below, 16 feet long, 11 feet in circumference at the thickest part, with 1 fin on the back. Head acuminated. Lower jaw blunt, longer than the upper, with two short lateral bony teeth. Upper jaw sharp, let into the lower one by two lateral impressions

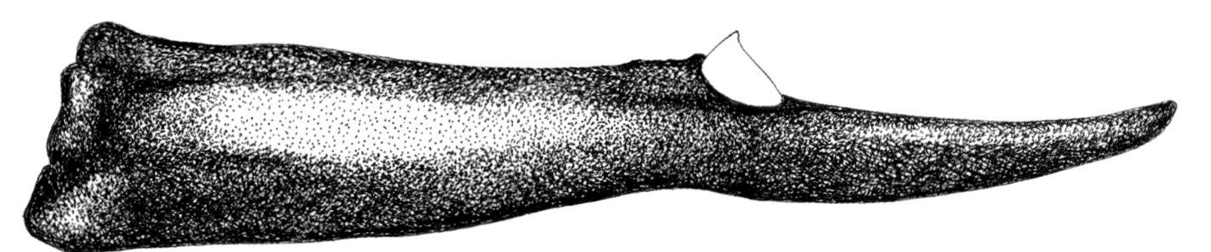

Mandible of *Mesoplodon bidens.*

> corresponding with the teeth. Opening of the mouth 1 foot 6 inches. Tongue smooth, vascular, small. Throat very vascular, rough. Under the throat are found two diverging furrows, terminating below the eyes; which are small, and placed 6 inches behind the mouth. Spiracle lunate, the ends pointing forwards.

Because it was the first *Mesoplodon* recognized by science, the original (type) specimen has been the subject of an unusual amount of historical and biological inquiry. In the *Archives of Natural History* for October 2013, British zoologist Geoffrey N. H. Waller recounted the history of Sowerby's "two-toothed cachalot," which (fortunately for science) was a male, so the unique nature of the teeth could be recognized. As Waller wrote, the skull itself had an interesting journey, as befitted an object of such importance: "After Sowerby's death in 1822, the skull was sold and bought by William Buckland who was closely associated with Oxford University and he donated it to the Anatomical Museum of Christ Church, Oxford University. In 1860 the skull was transferred to the Oxford Museum of Natural History, where it resides now as no. 06998 in the collection."

Tomilin (1967) says that Sowerby's beaked whale, now known as *Mesoplodon bidens*, reaches a confirmed length of 18 feet (5.5 m). Like the other Atlantic species of *Mesoplodon*, it has a streamlined body, small head with a pointed beak, and flukes without a terminal notch. Its color is dark gray to black above, lighter below, and the skin is often marked with white spots or scars. Photographs in the western North Atlantic sometimes show a light-colored phase.

The two teeth in the males (*bidens* actually means "two teeth") are typically flattened and are located at about the middle of the lower jaw. Unlike the symmetrical teeth of *M. mirus*, the teeth of this species are unusually shaped, having a long root set obliquely into the jaw and a crown that points toward the front. In females they usually do not erupt, and Fraser (1953) showed X-rays of vestigial small teeth in the lower jaws of a 12-foot male that stranded at Glamorgan, Scotland. For the most part, beaked whale identification relies upon dentition, so without a careful examination of the skull and teeth, it would be possible to confuse this species with almost any of the other beaked whales.

In 2004, Colin MacLeod and J. S. Herman published an extensive review of the "Development of tusks and associated structures in *Mesoplodon bidens*," for which he examined eighteen skeletal specimens from the collections of the National Museum of Scotland, half of which were males and therefore had a single erupted pair of teeth in the lower jaw. In this paper, MacLeod reiterated that the pattern of tooth-scarring in males "suggests that the tusks usually strike an opponent and move parallel to the longitudinal axis of its body, predominantly on the anterior portion of the dorsal surface. The development of the bony abutment behind the tusks is consistent with the hypothesis that the whale approach and pass each other dorsum to dorsum, attempting to inflict injury without being injured." "Dorsum to dorsum" means that one of the whales has to be swimming belly up (see illustration on p. 124), an arrangement not observed in beaked whales in the wild—nor, as a matter of fact, in any other mammals.

Even though there are no distinguishing characteristics that could serve as field marks at sea, in 1997, Carlström, Denkinger, Feddersen and Øien identified two adult beaked whales as Sowerby's beaked whales northwest of Norway, farther north than this species had ever been seen before. Carlström and her colleagues watched the animals swimming close to their vessel: "Their swimming behavior seemed calm and unhurried with a rolling motion . . . A number of morphological features were clearly seen during the encounter. The foreheads were slightly concave. A small projection near the middle of the lower jaw of one of the animals implied the presence of exterior teeth (an adult male character). Both animals were dark slate-grey with slightly paler sides and no visible scars. The dorsal fins were falcate and set well behind the center of the back." From this sighting, it became evident that this species, already the most northerly of the Atlantic beaked whales, may inhabit high polar latitudes. It has also been observed off the west coast of Norway, in the Baltic Sea, in the Mediterranean, and off the British Isles. There are a few records from the western North Atlantic, including Nantucket, Massachusetts, and Notre Dame Bay, Newfoundland. From the concentration of strandings, it appears that the population is centered in, but not restricted to, the North Sea.

Jonsgard and Hoidal (1957) studied stranded specimens on the west coast of Norway and discovered that mating usually takes place in the late winter and spring. After a gestation period of about a year, a single 7-foot (2.13 m) calf is born. During its first year the calf grows 3 to 4 feet, and it is weaned when it is about 10 to 11 feet (3 m) long. Generally regarded as a cold-water species, a single Sowerby's beaked whale was spotted stranding in shallow water near St. Joseph Spit on the Gulf coast of Florida in October 1984. Under veterinary care, it was taken to an isolated basin in an empty boathouse, but it died after three days. Postmortem examination revealed that it was an immature male, but no illness or cause of death could be identified (Bonde and O'Shea 1989). Besides this "extralimital wandering," other beaked whales known from the Gulf of Mexico are Gervais' (*M. europaeus*), Blainville's (*M. densirostris*), and Cuvier's (*Ziphius cavirostris*) (Jefferson and Baumgardner 1997).

According to Fraser's 1974 summary, 21 Sowerby's beaked whales had stranded on British coasts since 1913, and in 1977, Duguy reported: "This species also appears among those which are found occasionally on the Atlantic and Channel coasts." A rare opportunity to observe a juvenile occurred in 1972, when a calf was captured off Ostend, Belgium, where its mother was beached and dying. The baby, 8.8 feet (2.7 m) and later estimated to be less than a year old, was brought to the Delphinarium at Harderwijk, Holland, where it lived for three days before it died from a headlong crash into the tank wall. Postmortem examinations revealed a hydrodynamic shape designed for a species "built to go fast and straight, which seems logical for a pelagic animal. They are seriously hampered in narrow surroundings, let alone in a tank even when it is in a semi-circle with a radius of 15 m as at Harderwijk" (Dudok van Heel 1974).

Andrews' Beaked Whale

Mesoplodon bowdoini Andrews, 1908

Among the little-known beaked whales, Andrews' is among the least known. It has never been seen alive and is known from only a few dozen stranding records. The largest specimen was about 14 feet (4.4 m) long, so we assume that this species is one of the smaller of the beaked whales (Baker 1972). Because it appears to be the southern counterpart of a species that is known to be heavily scarred (Hubbs' beaked whale), it can be assumed that this species also is black with numerous white scratches and scars.

The type specimen, which originally came from New Zealand, was described by famed Gobi Desert dinosaur hunter Roy Chapman Andrews (1884–1960), who acquired the skeleton when he was Assistant Curator of Mammals at the American Museum of Natural History (AMNH) in New York. (He designated it as *Mesoplodon bowdoini,* in honor of George S. Bowdoin, a trustee of the American Museum, "through whose generosity the enlargement of the collection of Cetaceans in this museum has been made possible.") In the 1908 description in the *AMNH Bulletin,* Andrews described a tooth as "laterally compressed, convex posteriorly and slightly concave anteriorly; its surface is rugose. Its greatest height is 90 mm and its greatest length is 75 mm" (3.75 in high and 3 in long). In 2001, Alan Baker gave the maximum height of an adult male's tooth as 138 mm (5.30 in). In 1908, so little was known about the dental dimorphism in mesoplodonts that Andrews wrote, "The sex is unknown, but I believe it to be a male." He was correct.

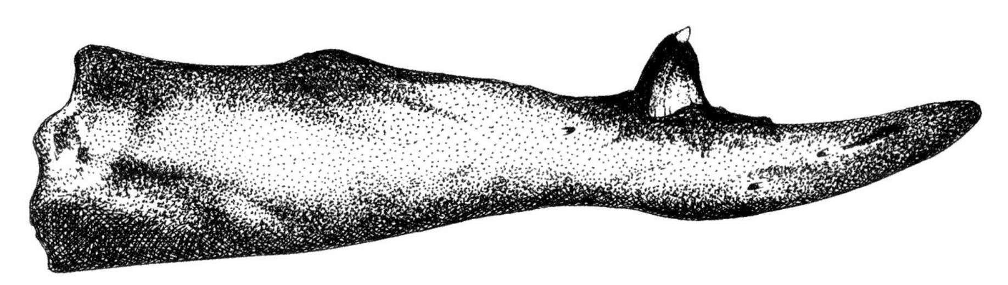

Mandible of *Mesoplodon bowdoini.*

As is the case with many other beaked whales, the males have two large flattened teeth set in partially raised sockets in the lower jaw. In *M. bowdoini*, these teeth often have a peculiar "hollow" on the anterior margin, the cause of which is unknown. Moore (1963) wrote that "the wear may have been accomplished by some kind of food that is long and slender and finely abrasive. . . . One could speculate that feeding on small, slender sharks with skin like fine sandpaper might abrade the teeth in this fashion." Not all specimens show this wear, indicating that it is an acquired characteristic, and it has also been shown that it is not restricted to this species. A Japanese specimen that Nishiwaki (1962a) identified as *bowdoini*—but cannot be, if we accept Moore's segregation of this species to the South Pacific—also had deep grooves in the teeth. That specimen that Nishiwaki (1962a) described is currently identified as *carlhubbsi*. Nothing is known of the life cycle of this species, but one female was found to be carrying a 5-foot (1.5 m) fetus in September, suggesting that the young are born in the southern spring (Gaskin 1968).

Because male *M. bowdoini*, *M. stejnegeri*, and *M. carlhubbsi* are all blackish, scarred animals with a white beak and prominent, flattened teeth, cetologists have had a particularly difficult time differentiating these varieties of beaked whales, occasionally even assigning them to the same species (Orr 1953). This is largely attributable to the limited amount of study material for comparative purposes and to the uncooperative nature of the beaked whales themselves. They do not strand very often because they are almost exclusively deepwater inhabitants of remote waters and they do frequent normal shipping lanes. When Alan Baker examined the "Status, Relationships, and Distribution of *Mesoplodon bowdoini*" in 2001, he wrote that "*Mesoplodon stejnegeri* is regarded as the most distinct species, as it is distinguished from *M. bowdoini* and *M. carlhubbsi* largely by the lack of prominential notches in the skull, and the shape and position of the male teeth." He describes the teeth of *M. bowdoini*: "In profile, the exposed part of the tooth is evenly convex posteriorly and slightly concave anteriorly, with a strong, blunt denticle projecting forward, upward and outward . . . The teeth are exposed only terminally, and in adult specimens show wear on the anterior inside edge and around the denticle. Such wear may result from rubbing against the upper jaw, and (the denticle) from contact with other males during agonistic sparring."

Up to 1988, all strandings of *Mesoplodon bowdoini* had been recorded from the Australasian region, but in that year the National Museum of New Zealand obtained two skulls from the Falkland Islands, and the South African Museum obtained two more from Tristan da Cunha, expanding the species' range into the South Atlantic. Natalie Goodall et al. reported a carcass of this species in Tierra del Fuego in 2004, and in 2005, Paula Laporta and three Uruguayan colleagues described a carcass found on the beach at La Coronilla, the first record of this species in Uruguay and the northernmost record in the Atlantic. In his 2001 review of *M. bowdoini*, Baker observed that this species has a circumpolar distribution north of the Antarctic Convergence at 32°S, but there is a gap from Chatham Island to the South American coast, probably a reflection of a general shortage of cetacean records for that part of the world. Andrews' beaked whale is now known from 37 records.

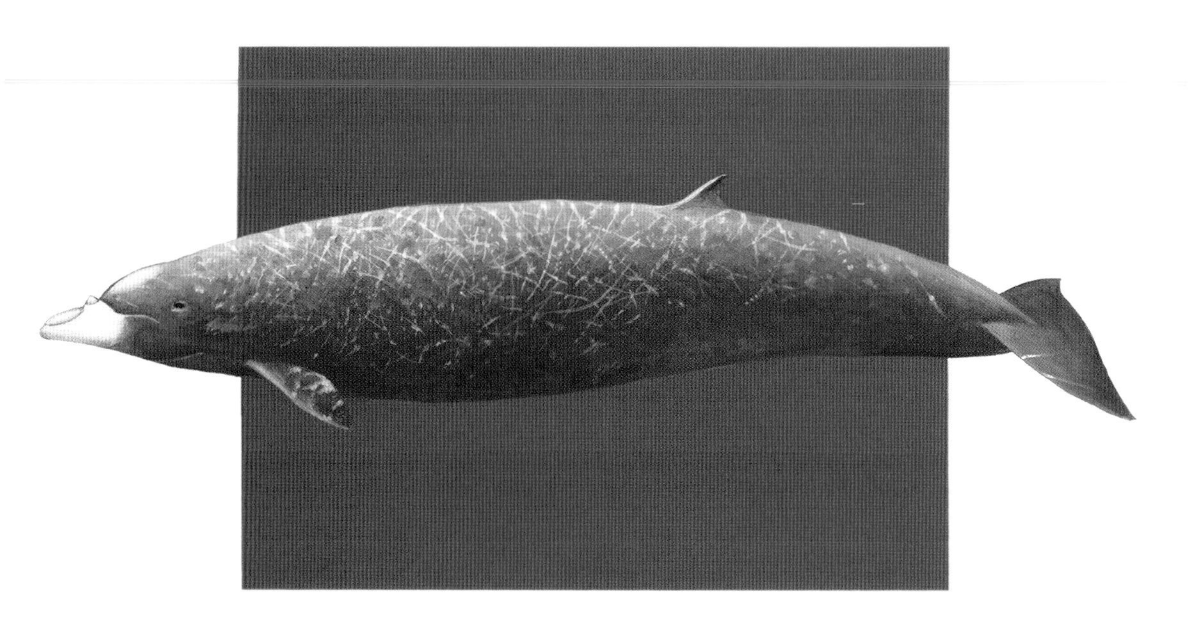

Hubbs' Beaked Whale

Mesoplodon carlhubbsi Moore, 1963

J. C. Moore described this species in 1963, making it one of the newer additions to the genus. He named the species after Carl L. Hubbs (1894–1979), one of the country's premier marine biologists. Hubbs (1946) described a 17-foot male beaked whale that stranded near his office at La Jolla, California, that Remington Kellogg of the Smithsonian had identified as *Mesoplodon bowdoini*. In 1953, R. T. Orr reviewed the species and concluded that *M. bowdoini* and *M. stejnegeri* "represent one species," the purported differences between them being attributable to variations in the age and sex of the animals described. He believed that the specimens previously assigned to *bowdoini* were old males of the genus *stejnegeri*. However, in his review of the North Pacific representatives of the genus *Mesoplodon*, Moore (1963) examined the skulls and teeth of the specimens that had been studied by Hubbs and by Orr. When he had separated the material by geographical location and cranial characteristics, he realized that *bowdoini* and *stejnegeri* were indeed distinct species. Furthermore, some of the skulls that had previously been assigned to these species were distinct enough from the others to warrant the establishment of yet a third species. Moore named it *carlhubbsi* after Hubbs, "the discoverer of the first good specimen, who has made other contributions to cetology, and whose name is, we believe, not entirely unknown among ichthyologists."

Hubbs' beaked whale reaches a length of 17 feet (5.3 m) and is built along the same lines as typical mesoplodonts: deep chest and belly, small head and tail. Adult males are dark gray to black, with no differentiation on the

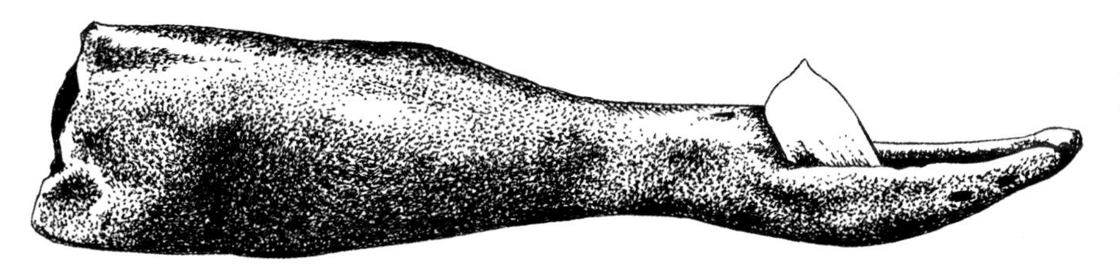

Mandible of *Mesoplodon carlhubbsi.*

underside, but females are white on the belly (Mead et al. 1982). Males and females both have a white beak, and the males have protruding teeth. Leatherwood and Reeves (1983) pointed out that "the most distinctive feature, found only in adult males, is the stark white convexity in front of the blowhole. It is reminiscent of a cap or beanie." As shown in photographs in Mead, Walker, and Houck's 1982 summary of the biology of *M. carlhubbsi,* adult males can be latticed with white scars caused by the protruding teeth of rival males. (No one has ever witnessed a tooth-raking of one male by another, but a profusion of parallel scars indicates that the males inflict these wounds with their mouths closed.) Those scars that are ovate, not linear, are likely the result of bites by the ubiquitous cookiecutter sharks. John Heyning (1984) based his theory of intraspecific fighting of beaked whales on a study of *M. carlhubbsi.* The scars on the bodies of the males, he wrote, "are undoubtedly caused by the aggressive use of teeth by other males of the same species. Although there is no observational data to substantiate either intraspecific or interspecific fighting, I suggest that interspecific aggression among male *Mesoplodon* occurs rarely, if at all." Male Hubbs' beaked whales inflict damage only on other male Hubbs' beaked whales, not on other mesoplodonts and not on females of their own species.

The large teeth are located about one-fifth of the distance from the apex to the posterior edge of the lower jaw. The teeth in males are more or less straight sided and have a point projecting forward. In this and the other saber-toothed whales, the large teeth are particularly thin and flat in cross section; one set of tooth dimensions was as follows: height, 6.3 inches (160 mm); width, 2.95 inches (75 mm); depth, 0.55 inches (15 mm), according to Nishiwaki and Kamiya (1959). From photographs that they published, it is clear that the lower teeth of adult males are covered with skin, except for the protruding tips. (This is also characteristic—if not diagnostic—of *Mesoplodon densirostris,* where only the tooth tips protrude from the dramatically arched jaw.) In other species, such as *M. stejnegeri,* illustrated in Nishimura and Nishiwaki (1964), the large teeth are fully exposed. Moore differentiated Hubbs' beaked whale from other species of *Mesoplodon* on seventeen diagnostic characters, concerned mostly with skull proportions, but also with the actual shape of the teeth. He described the appearance of the teeth: "From [the] front view the beak is seen to lie between the teeth like a zero between parentheses."

The known habitat of *M. carlhubbsi* is the temperate North Pacific Ocean. Specimens have been found between Prince Rupert, British Columbia, and San Diego, California, on North American coasts and at Ayukawa, Honshu, Japan (Mead et al. 1982). This species is now known from 53 records.

On August 24, 1989, two juveniles of *M. carlhubbsi* were found stranded alive on Ocean Beach, San Francisco. They were "rescued" by staff of the Marine Mammal Rescue Center and taken to Marine World Africa USA, a (now defunct) oceanarium complex in Vallejo, California. Named "Nicholas" and "Alexander," the two whales, each about 9 feet long, were placed in a shallow, 30-foot-diameter pool and fed a high-calorie formula four times a day. Before they died, two weeks after their "rescue," at least one of them (Alexander) was heard (and recorded) to produce "two hundred ninety six pulse sequences and six whistles" (Lynne and Reiss 1992). It is not known if the sounds made by these young animals were intended for echolocation or communication.

Blainville's Beaked Whale (Dense-Beaked Whale)

Mesoplodon densirostris (de Blainville, 1817)

In adult males of this species, the modified lower jaw is a prominent and noteworthy characteristic. It is highly arched, and the teeth "protrude above the forehead somewhat like a pair of horns" (Rice 1978). The jaws are also unique in their configuration and serve to accommodate the massive teeth. The beaked whales with exaggerated teeth are known collectively as the "deep-socketed whales" (Moore 1968), since only a small portion of the teeth of the mature male protrudes beyond the jawbone. (Mörzer Bruyns [1971] referred to these species as the "saber-toothed whales.") Harmer (1927) described the jaw thus: "The lower jaw has a remarkable shape, due to its extraordinary depth in the region of the teeth, which are situated near the middle of the jaw. From this point the dorsal outline of the bone slopes down steeply to the low anterior region which corresponds with the beak above it." Out of the socket, one tooth was measured at 6 inches (152 mm) in height, 3.375 inches (85.9 mm long, and 1.75 inches (44.5 mm) deep (Norman and Fraser 1938). Adult females and juveniles have the same jaw configuration, but the teeth, embedded in the gums, are not visible. Only *Mesoplodon stejnegeri*, *M. carlhubbsi*, and *M. bowdoini* have teeth that are as large as those of the dense-beaked whale (the straplike teeth of the male *M. layardii* and *M. traversii* serve a completely different function, whatever it might be), but no other species has the high, crested

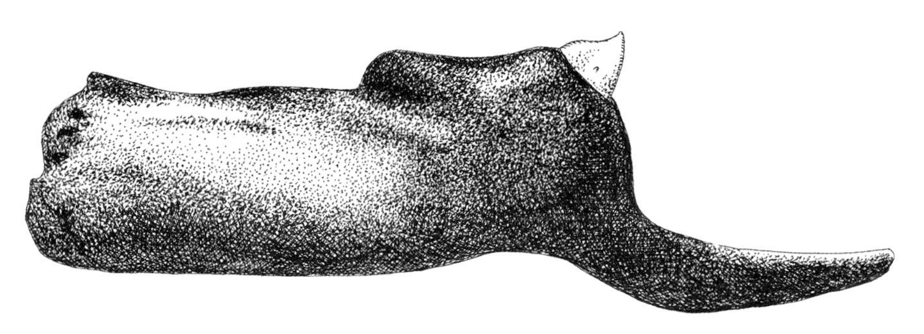

Mandible of *Mesoplodon densirostris.*

mandible. Moreover, the teeth of the other species are shaped differently, and the skulls have measurable proportional differences.

The dense-beaked whale is dark gray to black in color, usually with white scars and blotches. As with most beaked whales, there is no notch between the left and right lobes of the flukes, and the flippers and dorsal fin are proportionally small. The maximum recorded length for this species is 15.6 feet (4.0 m), and a healthy animal this size would weigh around 1,000 kg, or a little over a ton. At sea it would be almost impossible to identify this species unless one could get a close look at an adult male. (This species is encountered in tropical waters more frequently than other mesoplodonts, but not enough specimens have been seen to make a convincing argument for the animal's distribution. As of 1971, Besharse listed all eighteen specimens known to date. When he described one of the dense-beaked whales from Midway, it was the only adult female then known to science. Rice (1978) provided one of the earliest published observations of living dense-beaked whales seen at sea: "I once saw a group of ten or twelve in the Hawaiian Islands. They blew leisurely at the surface for several minutes, and then dived. We waited forty-five minutes for them to come up, but we never saw them again." On Midway Island, Galbreath (1963) discovered three beaked whales stranded at the same time. Two were dense-beaked whales, and the third was a goosebeak whale. More recently, it has been shown that these two species are often seen in the same neighborhood, particularly around the Big Island of Hawaii (McSweeney et al. 2007).

Occurring from tropical to temperate waters of the Atlantic, Pacific and Indian Oceans, *M. densirostris* has the widest distribution of any *Mesoplodon* species, but it is nowhere common. It has been recorded from the tropical and warm temperate waters of the world's oceans. (Rice [1977] offered "tropical beaked whale" as an alternative common name for the species.) Strandings include such diverse locations as New Jersey, Madeira, South Africa, Canada, Massachusetts, and the Bahamas in the Atlantic; Lord Howe Island and Queensland in Australia; Midway Island; Algoa Bay, South Africa; Tasmania; and the Seychelles. In 1968, two females were harpooned off Formosa and brought to the fish markets (Kasuya and Nishiwaki 1971). For the North Atlantic population, Moore (1966) has suggested a more offshore distribution than that of the other beaked whales of that region. There are nine European records, including the type, and seven South Atlantic records. We now have 96 records of this species.

For all its wide distribution, however, Blainville's beaked whale is still a very rare animal indeed. In April 2010, when a beaked whale stranded in Subic Bay, on the Philippine island of Luzon, a local newspaper ran this headline: "Elusive Beaked Whale Stranded in Subic." Henry Empeño wrote, "The whale, a male specimen of the Blainville's beaked whale, was seen circling for two days, then ended up dead at the seashore. . . . Dr. Leo Suarez, a marine biologist at the Ocean Adventure, said this was the first time for biologists to document the beaching of a Blainville's Beaked Whale in the country." There are probably more where that one came from.

Henri Marie Ducrotay de Blainville described this species in 1817 based on a 7-inch-long piece of upper jaw bone that he found had a higher density than elephant ivory, inspiring the species name densirostris, from the Latin *densum rostrum*. Other beaked whales show a similar increase in density of the rostral bones, but none to the extent of *M. densirostris*. The bone that makes up the rostrum, wrote de Buffrénil, Zylberberg, Traub, and Casinos in 2000, "has the highest density, mineral content, and compactness of any bones hitherto observed for mammalian bone." The density of the rostral bones of this species—more pronounced in males than in females—has become a subject of discussion, conjecture, and conflict among those who would study beaked whales. Li and Pasteris (2014a,b), mineralogists at Washington University in St. Louis, have been studying the hyper-mineralized rostrum of *M. densirostris* as an example of bone growth. There is no question that they are the densest bones ever examined, denser than elephant ivory or the rocklike tympanic bulla (earbone) of the cetaceans, but why did this develop, and what does it do for the whale?

In a 1984 discussion, John Heyning suggested that rostral bone density (in the very similar *M. carlhubbsi*) was a mechanical reinforcement during intraspecific fights of adult males: "The densely ossified mesorostral canal evident in adult males probably functions in reinforcing the rostrum as the males engage in fighting." This interpretation seemed obvious and even logical, but examination of the rostral bones (in *M. densirostris*) with a scanning electron microscope by Zioupus, Currey, Casinos, and de Buffrénil (1997) found that the bones forming the rostrum (premaxillae, maxillae, pterygoids, palatines, and vomer) are extremely brittle and unbending, characteristics that, "Far from improving resistance to fracture . . . facilitates rapid and extensive propagation of cracks. This poor mechanical behavior is a general characteristic of very compact and hyper-mineralized bones. Bones adapted for shock loading, such as deer antlers, have the opposite structural characteristic."

In 2000, de Buffrénil et al. wrote that these bones are particularly susceptible to fracture and therefore, "Heyning's interpretation is discounted, being most improbable on mechanical grounds." (In a 2010 study, Cozzi and colleagues wrote, "The aggressive interactions in Ziphiidae have not yet been observed and can only be inferred analyzing the external scarring patterns on the bodies of sighted or stranded specimens.") As alternative explanations, Zioupus et al. suggest a possible hydrostatic role, where the heavy bones act as ballast to "reduce the energetic cost of diving," but no mention is made of having to carry these heavy bones upward on the ascent. Why then the bone density? "Very compact, highly mineralized bones are well suited for ultrasonic signal construction," wrote de Buffrénil et al., and the dense bones, along with the unusual sagittal crests of the skull, might be used for echolocation of prey items in the depths.

Zioupos et al. (1997) also concluded that the rostrum of *M. densirostris* was too brittle to function effectively to prevent damage during male-male combat. They concluded that the mesorostral ossification was well adapted for the transmission of sound, owing to its extreme density and therefore that the dense rostrum of *M. densirostris* might serve some acoustic function.

MacLeod (1998) took the discussion further into the realm of speculation, however, as he attempted to explain the curious pattern of scars on *M. densirostris*. He reproduces a photograph of an adult male, showing an abundance of scars on the dorsal surface around and aft of the blowhole, aligned along the whale's long axis. Then we get a drawing, showing what he believes is the sequence of combat of adult males, where they approach each other head on, but one of them is upside down. This puts the fighting tusks of the whales in position to rake one another around the head and back, and he suggests that this longitudinal movement, rather than head-on collisions, explains the structure of the rostral bones: "The longitudinal grain of the structure along which the fractures spread would provide more, not less, protection against severe fractures, as the grain of the rostrum would prevent fractures passing vertically through the rostrum, or transversely across it, breaking it in two." MacLeod may have solved the problems of bone density, linear scarring, and male fighting, but he raises additional questions: if two males are approaching each other to do battle, how do they decide which one is to turn upside down? (Does the aggressor issue a challenge by approaching upside down?)

During his field work in the Bahamas, MacLeod observed that this species tended to congregate in groups, consisting of a number of females, calves, and/or juvenile animals, often accompanied by a single male with fully erupted tusks and perhaps a single prepubescent male. Single animals were observed twice; one was a pubescent male and one an adult male (MacLeod and D'Amico 2006). Lone males or males with female groups may be related to this species' fighting inclinations.

In his 2009 discussion of beaked whale research in Hawaii, Robin Baird of the Cascadia Research Collective in Olympia, Washington, reported that he had spent more than 300 days at sea looking for (and finding) mesoplodonts. The species they encountered most frequently were Cuvier's and Blainville's. They were uncommon, but Baird and his colleagues persevered, and Baird wrote, "A collaboration with Dan McSweeney, a researcher working off the island since the mid-1980s has played a large role in this—while working with other species, Dan has been taking beaked whale photographs opportunistically since the mid-1980s, and combined with our photos this has resulted in the longest-term photo-identification catalogs available for either species world-wide, allowing us to examine social organization and long-term site fidelity."

A dense-beaked whale, which stranded in 1969 at Crescent Beach, Florida, and lived for only 24 hours at Marineland of Florida, was heard to emit "chirps or short whistles," which were later analyzed as pulsed sounds by Caldwell and Caldwell (1971). From those humble beginnings, an entire discipline has arisen; recording and analyzing the sounds made by beaked whales—particularly *M. densirostris*—has become one of the cornerstones of beaked whale research. "Tagged whales," wrote Tyack, Johnson, Madsen, et al. (2008),

> are usually silent when starting a dive, but start producing echolocation clicks at a few hundred meters depth, shallower than the depth at which they feed, suggesting that descending whales can scan the deep layers where they will feed. Once sperm or beaked whales encounter prey, they switch from regular search clicks to a buzz of rapid clicks. Tags on beaked whales not only record outgoing clicks, but also echoes from prey at ranges out to 10–20 m. Beaked whales produce clicks every 0.2–0.4 sec. when searching. Beaked whales will pass by many targets before selecting one. Whales may switch from the search clicks to a buzz as they close within a body length of the prey.

Like all the other beaked whales, *M. densirostris* is not common, but its occasional onshore and offshore visibility and availability has made this species the "poster child" for beaked whale studies. Their echolocation behavior has been documented in the Bahamas; Robin Baird and his associates from the Cascadia Research Cooperative have been following, tagging, and photographing this species off Hawaii since 1999; and the large nearshore population off the western Canaries has provided an unprecedented opportunity to listen to what the whales were talking about. In 2004, Johnson, Madsen, Zimmer, Aguilar de Soto, and Tyack published "Beaked Whales Echolocate on Prey" in the *Proceedings of the Royal Society*, and the same authors wrote refinements of that article in 2005 and 2006 in the *Journal of Experimental Biology*, demonstrating the undeniable usefulness of these whales in our attempts at understanding the complexities of the acoustic world of the beaked whales.

If you want to see dense-beaked whales up close and personal, the Canary Islands in the eastern North Atlantic, off the coast of West Africa, would seem to be among the best places to go. In order to observe this species at the surface and underwater from whale-watching vessels off the island of La Gomera, Fabian Ritter and Bernd Brederlau (1999) made more than 500 whale-watching trips, accumulating nearly 1,800 hours of recorded sightings. Sometimes the whales approached the boats; sometimes they avoided them. The researchers observed that "the colour patterns of animals in several different groups were striking, as some of the animals had ochre to gold fins, as well as coloured patches in the eye-, flipper-, and blowhole-area." (These patches are probably diatom infestations and do not represent the actual coloration of the animal.) "Another sighting," wrote Ritter and Brederlau, "revealed a surprisingly active behavior: a group of two adults and a juvenile sprinted several hundred meters with the animals repeatedly porpoising at high speed." The investigators also jumped in the water several times to see how the whales would react to swimmers joining them: "At times they showed very curious behaviors towards swimmers, e.g., approaching, scouting, and parallel swimming. During the longest sighting (1 h 40 min, group size 9 animals) [they] were able to take underwater pictures of each member of the group, including females, calves, juveniles, and a mature male."

For reasons that probably have to do with prey availability in the vicinity of deepwater seamounts, a population of *M. densirostris* lives year round close to the shore of El Hierro, the westernmost of the Canaries, which makes them comfortably accessible to researchers who would study their habits. From 2003 to 2010, scientists tagged fourteen individual beaked whales with suction-cup-attached digital recorders to collect information on the diving patterns of the whales as they foraged in the depths. Various researchers have analyzed the click sounds of these accessible subjects, producing a more or less comprehensive picture of what the whales are doing, and how they're doing it. For example, in the 2011 report "Following a Foraging Fish-Finder," Arranz, de Soto, et al. tracked the whales' deep-diving activities, echolocation sounds, and the "distinctive buzzes indicating prey capture attempts." They wrote, "Compared to sperm whales, another echolocating predator at similar depths, Blainville's beaked whales spend noticeably less time per dive searching for prey." Their analysis of *M. densirostris'* foraging concludes with:

> We have shown that Blainville's beaked whales spend only four hours per day hunting for food, with such a short foraging time likely resulting from long transport times to foraging depths and long recovery times between

> deep foraging dives. This necessitates a stable and abundant prey resource that can be located reliably in an extensive 3-dimensional world of darkness. The steep sloping terrains in locations where Blainville's beaked whales are often found may offer access to resources associated with both the DSL [Deep-Scattering Layer] and the BBL [Benthic Boundary Layer] over a small spatial scale. Echolocating whales can glean both biotic and abiotic cues to aid the efficient location of these resources from biosonar echoes. The enigma of why Blainville's beaked whale abundance is apparently strongly linked to a dense DSL, even though they seem to forage outside the DSL, may then be explained by the observation here that whales target prey in the oxygen minimum layer associated with, but deeper than the bulk of the DSL. Thus, by inhabiting steep undersea slopes, Blainville's beaked whales can target a stable and abundant resource of mixed meso and benthopelagic prey using biosonar-derived landmarks.

In 2006 and 2008, scientists affixed satellite tags to eight individual Blainville's beaked whales off the Big Island of Hawaii. Schorr, Baird, and their colleagues (2009) reported that the whales did not wander far from the original tagging locations, demonstrating an unexpected site fidelity, the inclination not to wander far from their breeding areas. The 2006 report by many of the same scientists (Baird et al.) concentrated on the diving behavior of these whales and Cuvier's beaked whales (*Ziphius cavirostris*), often found in widely spaced mixed groups in the deep waters off the west coast of the island. *Ziphius* was recorded to dive deeper—indeed, this species now holds the record for the deepest dives ever recorded for any cetacean—but *M. densirostris* proved no slouch, recording several dives in the neighborhood of 1,400 meters (4,992 ft), and one dive in which an adult female reached a depth of 800 meters (2,600 ft), accompanied all the way down (and up) by a calf. There was no significant difference in depth or duration between day and night dives, suggesting that their prey (squid for the most part) do not make daily vertical migrations.

It is almost a given that beaked whales inhabit deep waters, but different species seem to prefer different locations and different topography. In their 2005 study of habitat utilization by Blainville's beaked whales off Great Abaco Island in the Bahamas, MacLeod and Zuur wrote that they might frequent the deeply sloping regions around seamounts because certain cephalopods "become concentrated in the near-bottom layer and so form an important resource of predators." Similar seamount topography has been identified in other regions where *M. densirostris* has been observed, such as the Canaries, Hawaii, and the Society Islands.

Gervais' Beaked Whale

Mesoplodon europaeus (Gervais, 1855)

In 1959, a beaked whale washed ashore at Boca Grande, Florida, and Eugenie Clark, director of the Mote Marine Lab in Sarasota, was called in to examine it. She realized that it must be a species of *Mesoplodon*, but since the teeth and lower jaw were missing, she was unable to identify it. After a careful search—much of it conducted offshore and under water—she found the lower jaw and sent a sketch of it to J. C. Moore, the authority on beaked whales, then at the American Museum of Natural History in New York. He identified it as *Mesoplodon gervaisi* but he was distressed about the missing teeth, since this was only the second male ever found (in the lower jaw of the male, the sockets for the teeth are plainly visible). The son of the fisherman who had discovered the beached whale had taken the striking, ivory-like teeth home for souvenirs, but in the name of science he voluntarily surrendered them to Clark, who forwarded them to Moore, who confirmed it was *M. gervaisi* (now known as *M. europaeus*).

The species, originally named *Dioplodon europaeus*, was described in 1855 by Paul F. L. Gervais (1816–79) a French zoologist and anatomist, who, with Pierre Joseph van Beneden (1809–94), published the classic atlas of cetacean osteology that is still in use today (Beneden & Gervais 1868–79). *M. europaeus* reaches a maximum known length of 16 feet (4.9 m), but one individual was reported as 22 feet (6.7 m) long. The head is small, the beak narrow, and the flukes wide, measuring as much as 33% of the total length of the whale. Normal coloration is black or dark gray above, sometimes with a white patch on the undersides. The dorsal fin, like that of many members of the genus *Mesoplodon*, is

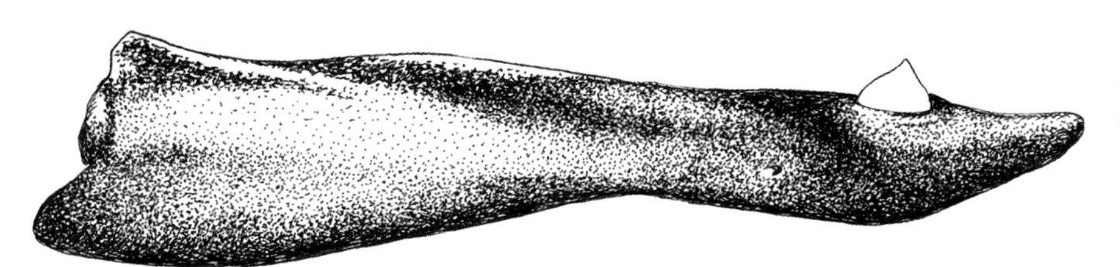

Mandible of *Mesoplodon europaeus.*

small, slightly falcate, and located far posterior to the midpoint of the back. Only adult males have erupted teeth, which appear about one-third of the distance from the tip of the jaw to the corner of the mouth, not quite at the midpoint of the lower jaw. As with males of the other *Mesoplodon* species, the teeth protrude prominently outside the upper jaw. In his description of a specimen stranded in Cuba, Varona (1970) illustrated the teeth of an adult male and showed them to be roughly triangular in shape and flattened in what Moore (1968) referred to as the "antero-posterior plane"—in other words, they are flattened along the animal's long axis. In Varona's specimen, a tooth was measured at 2.75 inches (68 mm) high, 1.75 inches (42 mm) long, and less than 0.5 inches (11 mm) deep.

In 1953, two specimens came ashore at Bull Bay, Jamaica: a 14-foot lactating female and a 7-foot calf. Rankin's article about the pair (1955) included "the first set of photographs ever to be published of the external appearance of this whale" and described the female as follows: "The shape of this whale is rather elegant, I think, with its tapering snout and dome-like crown of the head and streamlined body demarcated from the head by a distinct "neck" and ending posteriorly in horizontal and backswept tail flukes. Even the blow-hole is beautifully curved and is situated on top of the head as far back as the eyes. The latter features are somewhat small and beady and may be easily overlooked, especially if the lids are shut."

In October 1977, a 17-foot (5.2 m) female stranded in Florida. She was accompanied by a 6.7-foot female calf that was kept alive for a few days at Sea World, Orlando, Florida. The female was black, with white on the tips of the upper and lower jaws, and the calf was a soft gray, darker on the back and on the tip of the lower jaw. The calf was marked with vertical creases, sometimes referred to as "fetal folds," which indicate that it was curled up in the uterus shortly before it stranded with its mother.

Despite its name, *Mesoplodon europaeus* is known from the coasts of North America, the islands of the Caribbean, West Africa, and Europe. It has been recorded from New Jersey, Long Island, North Carolina, Florida, Texas, Cuba, Jamaica, and Trinidad, but the first known specimen was discovered floating in the English Channel in 1840. Because this specimen was unique, "much doubt has been thrown on the validity of the species" (True 1910), but when two more specimens appeared on the shores of New Jersey in 1889 and 1905, the identification was confirmed. Until 1980, the English Channel specimen was considered a stray, but when a carcass showed up on a beach in Guinea-Bissau (West Africa) and another on a beach in Mauritania (also West Africa), it was clear that the English Channel stranding was not as much an aberration as had been earlier suspected (Robineau and Vely 1993).

Other reports have come in from Ireland, the Canary Islands, and Portugal, demonstrating—albeit spottily—that the distribution of *M. europaeus* spans the North Atlantic. As an indication of the ever-widening range of this species, lists of the common names for *M. europaeus* include "Gervais' beaked whale," "Gulf Stream beaked whale," "Antillean beaked whale," and "European beaked whale" (Mead 1989c).

Best et al. (2007: 125) recorded strandings from Ascension Island, Namibia, and Brazil in the South Atlantic. The Smithsonian has specimens that stranded on Ascension Island that have teeth differing slightly from the teeth of specimens from the North Atlantic; they may represent a separate subspecies.

We have come a long way since 1976, when Reeves and Ulmer, discussing *M. europaeus, M. mirus,* and *M. densirostris* that stranded on New Jersey beaches, wrote, "The genus *Mesoplodon* is a taxonomic nightmare. Twelve species are currently recognized, three of which have occurred in the New York Bight. Several *Mesoplodon* species have been described from only a few stranded carcasses, for some species sighting records at sea are non-existent. *Mesoplodon* species are considered offshore forms whose populations are probably small. No *Mesoplodon* species is known to have been extensively hunted, so these whales' perceived rarity is probably due either to natural inhibitions limiting their populations or to discrete behavior that prevents people from seeing them."

Their ranges overlap and they are roughly similar in size, shape, and coloration, so it would be easy to confuse *M. europaeus* with *M. mirus* (True's beaked whale). In fact, J. C. Moore and F. G. Wood published a paper in 1957 that they called "Differences between the Beaked Whales *Mesoplodon mirus* and *Mesoplodon gervaisi*." The teeth in adult males are different—at the tip of the mandible in True's and in the middle of the lower jaw in Gervais'—but not all specimens are adult males. As J. C. Moore pointed out in 1968, a peculiarity of the male Gervais' is that the two teeth protrude outside the mouth but fit into grooves in the skin of the outer upper jaw, a characteristic that has never before been observed in this—or any other—species of *Mesoplodon.* There are, say Moore and Wood, nine comparative measurements of the skulls, "which, used collectively, will separate the two species," and "the length of the flipper of *mirus* generally exceeds that of *gervaisi* in proportion to total body length." In other words, the position of the teeth can de used to differentiate adult males; for anything else, call your local beaked whale authority.

We now have 151 specimens and sightings from the North Atlantic and 6 from the South Atlantic. Podesta et al. (2005) described the only known record from the Mediterranean, an adult skeleton in the Museum of Natural History in Milan that stranded near Castiglioncella, Italy.

Ginkgo-Toothed Whale

Mesoplodon ginkgodens Nishiwaki and Kamiya, 1958

In September 1957, a live beaked whale came close to shore at Oiso Beach, near Tokyo, where some boys were playing ball. The boys waded into the water and killed the whale with their baseball bats. When Nishiwaki and Kamiya (1958) examined the specimen, they realized that its measurements and proportions did not agree with those of any known species of *Mesoplodon,* and they therefore "ventured to settle a new species for this specimen and designated it as *Mesoplodon ginkgodens.* This species name is chosen from the fact that the lateral view of the teeth of the present specimen resembles closely the shape of a leaf of the ginkgo tree (*Ginkgo biloba* Linneaus)." The teeth are longer than they are high, 4 inches by 2.5 inches (115 mm by 65 mm), compressed laterally, and located about one-third of the distance from the tip of the lower jaw to the other end of the jaw. As in other species of *Mesoplodon,* the teeth erupt over the gums only in the males.

The color of this whale is blackish gray, but it bears very few of the white scars that are so characteristic of all *Mesoplodon* species. In a 1972 paper, Nishiwaki and associates commented on this phenomenon with regard to an old male that had stranded near Ito City, Shizuoka Prefecture, in 1971. This animal had very few scars, "perhaps due to the inerupted teeth" of this species. Since so little of the tooth of this species is exposed, it would be difficult for these animals to inflict scratches on other individuals.

From the various Japanese specimens that have been carefully measured, the maximum length has been estimated at 17.3 feet (5.3 m). The flukes are

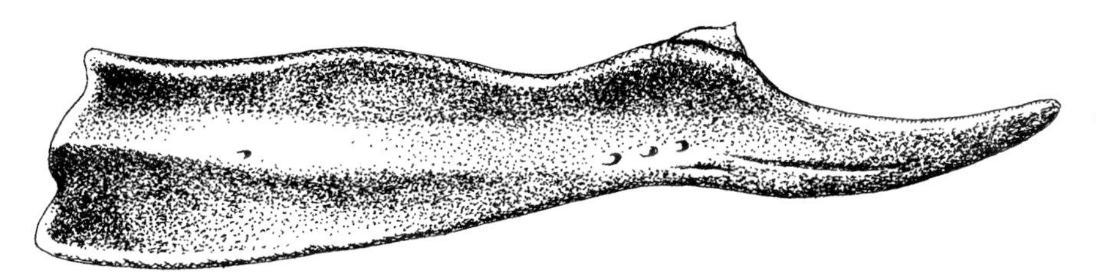

Mandible of *Mesoplodon ginkgodens.*

very wide, as much as 25% of the body length, and they show a slight unnotched convexity. The dorsal fin is located approximately two-thirds of the way down the back and is markedly falcate (Nishiwaki et al. 1972).

Of the eight ginkgo-toothed whales stranded on Japanese and Taiwanese beaches, six were discovered between 1962 and 1972. When this was established as a valid species, other previously unidentifiable specimens could be referred to it. The first specimen of *Mesoplodon* from Japan, reported in 1935, was variously identified as *M. densirostris* or *M. bidens*. From the shape of the teeth and other characters, Nishiwaki and Kamiya (1958) were "of the opinion that the first specimen might belong to the same species as the present specimen, viz. *Mesoplodon ginkgodens*." In 1954, a 16-foot female drifted ashore at Del Mar, California, 15 miles north of San Diego. It was not identified until 1965, when J. C. Moore, revising the genus *Mesoplodon*, examined this specimen, using the data and photographs published by Nishiwaki and Kamiya in their 1958 description of the new species, and incorporated it into his 1968 revision of the relationships of the beaked whales. Deraniyagala (1963a) published a description of "a new beaked whale, *Mesoplodon hotaula*" from Ceylon, but Moore examined this specimen, too, and said it was just another *M. ginkgodens* (Moore and Gilmore 1965).

But Deraniyagala's "new" beaked whale would not go away. Based on DNA analysis and physical evidence, the name *hotaula* was reapplied to several specimens that closely resembled *ginkgodens*, except that the males' leaf-shaped teeth were taller than they were wide, the opposite of the ginkgo tooth. Either a new species or a subspecies, these beaked whales were named *Mesoplodon hotaula* or *Mesoplodon ginkgodens hotaula* (Dalebout et al.

Leaf of the ginkgo tree, *Ginkgo biloba*.

2012). Because of this revision, discussions of the range of *M. ginkgodens* often conflated the two species (or the species and the subspecies), so we must subtract known *hotaulas* from any summaries that predate the Dalebout et al. paper. The known range of *M. ginkgodens*, therefore, is the waters of Japan (seven specimens); Taiwan (four specimens); southern California (one specimen), the Caroline Islands, the Galápagos Islands, and New South Wales (and maybe the Philippines and New Zealand).

In December 2012, a 16-foot-long beaked whale beached itself and died in Maco, in the Compostela Valley province, on the Philippine island of Mindanao. It was identified in the *Philippine Daily Inquirer* as "a super rare ginkgo-toothed beaked whale . . . only the second of its kind seen in the Philippines since 1957" (an unlikely date: Nishiwaki and Kamiya's original description of *M. ginkgodens* was based on a whale that washed ashore in Japan in 1957). The whale was said to have died of indigestion because of "garbage debris in its stomach." The photograph that accompanied the December 23 article certainly showed a beaked whale, but with no visible teeth (necessary for a positive identification), so calling this specimen *M. ginkgodens* is, at best, questionable.

Seamounts are mountains that rise from the seafloor but do not reach the surface; if they did, they would be islands. They are typically formed from extinct volcanoes and can be anywhere from 2,000 to 13,000 feet high, with peaks often thousands of feet below the surface. Largely unstudied, seamounts are believed to have significant effects on their immediate environments, affecting ocean currents and enhancing biological productivity, therefore attracting predators to the vicinity. From April to November 2005, researchers Johnston, McDonald, Polovina, Domokos, Wiggins, and Hildebrand (2008) deployed acoustic monitors at a depth of around 1,200 feet at the Cross Seamount about 180 miles south of Oahu. They were expecting to hear the sounds made by Blainville's and Cuvier's beaked whales, the species most often encountered in Hawaiian waters (see Baird et al. 2006), but they recorded sounds that they believed were "produced by a species of beaked whale other than Cuvier's or Blainville's." In a letter to Ellis in April 2013, Mark McDonald wrote, "We now suspect these are Gingko toothed beaked whales though we see this signature in echolocation signals as far as the Gulf of Mexico, where ginkgo toothed whales are not yet known to occur. These signals are the highest frequency of any of the beaked whales and appear to have the shortest detection range, probably only 200 m."

Palmyra Atoll is part of the northern Line Islands, an unoccupied U.S. territory in the central Pacific, several hundred miles south of the Cross

Seamount. In field seasons in 2007 and 2008, Baumann-Pickering, Wiggins, Roth, Roch, Schnitzler, and Hildebrand (2010) recorded "upsweep frequency modulated pulses reminiscent of those produced by beaked whales [but] these signals had higher frequencies, broader bandwidths, longer pulse durations and shorter pulse intervals than previously described pulses of Blainville's, Cuvier's, and Gervais' beaked whales. They were distinctly different temporally and spectrally from the unknown beaked whale at Cross Seamount."

Off Palmyra Atoll, Baumann-Pickering et al. heard sounds that were distinctly different from any they had heard before; "genetics on beaked whale specimens found at Palmyra Atoll suggest the presence of a poorly known beaked whale species." As noted above, using phylogenetic analysis of mtDNA, nuclear gene sequences, and morphological data, Merel Dalebout and her colleagues (2012) published "genetic and morphological data supporting the recognition of a previously described but unrecognized *Mesoplodon* beaked whale in the tropical Indo-Pacific. Currently known from at least seven specimens (Sri Lanka [1], Kiribati [1+] Hawai'i [3], Maldives [1], Seychelles [1]), this whale is the sister-taxon to *M. ginkgodens* proper" (the three "Hawai'i" specimens were from Palmyra Atoll). The type specimen (Sri Lanka) was described as a new species, *M. hotaula*, in 1963 by Deraniyagala. Dalebout, Baker, et al. (2012) argued "that *M. hotaula* represents a new subspecies (*M. ginkgodens hotaula*), if not a distinct species (*M. hotaula* Deraniyagala, 1963) in its own right." The mandibular teeth of *M. ginkgodens* are wider than they are tall, but the opposite is true for *M. hotaula*.

Unique among the currently recognized species of beaked whales, male *M. ginkgodens* do not become heavily scarred because so little of the teeth protrude. The ginkgo-toothed whale is found only in the Pacific, but "*M. hotaula*" is found in the tropical Indian and Pacific Oceans.

Will a completely new *Mesoplodon* species identify itself by its upsweeps? Maybe a trip to Cross Seamount would reveal the answer.

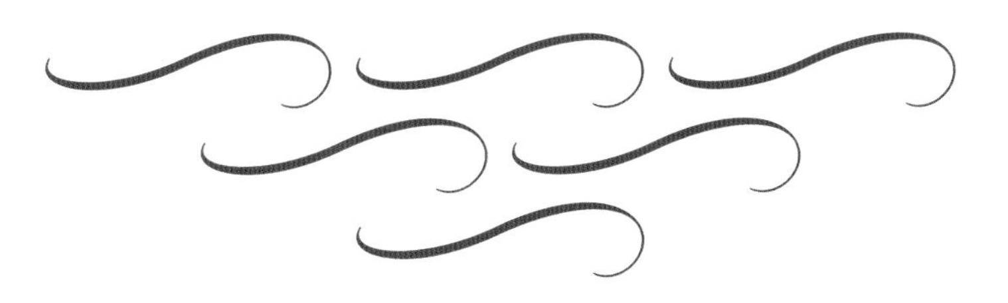

Gray's Beaked Whale

Mesoplodon grayi von Haast, 1876

Although this species is best known from stranding records in the Southern Hemisphere—particularly New Zealand—one of the best illustrations and descriptions are of an animal that beached in the totally unexpected location of Kijkduin, in the Netherlands: "[The] back is black to slate grey on the sides. Toward the ventral surface the color gradually becomes lighter, the sides being of a brownish grey. The ventral surface is of a light grey with a brownish tinge with the exception of a broad darker median band gradually becoming mottled anteriorly and vanishing in the region of the flippers. The flippers and tall flukes on both sides are very dark grey to black, the edges of the flippers have a light border (Boschma 1950)."

The head tapers to a long pointed beak without much of a forehead bulge, and in adult animals the head and beak are often white. The flippers are small and the flukes broad, without a median notch. As with most of the other beaked whales, males are often marked with scars and scratches. The maximum known length for a male is 20 feet (6.0 m); for a female it is 19 feet (5.8 m).

The two onion-shaped teeth, flattened laterally, are located about mid-way in the lower jaw and erupt only in adult males. The species is further characterized by the presence of seventeen to nineteen pairs of upper teeth, in addition to the lower pair. Boschma (1951a) reported that these minute teeth (none of which are longer than 0.375 inches [10 mm]) are present in all examined specimens, and "they constantly form a so distinct regular row that

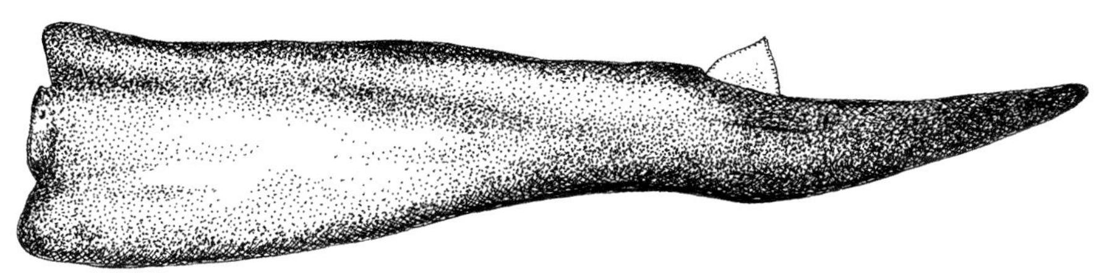

Mandible of *Mesoplodon grayi.*

they seem to perform a distinct function." Frank Robson (1975) wrote that "these teeth are so small that they are generally overlooked."

Gray's beaked whale has a circumpolar distribution in the Southern Hemisphere, south of 30°S latitude. It has been recorded in the Indian Ocean (South Africa); off South Australia and Victoria; and from New Zealand to the coast of Peru and Chile in the South Pacific. In the South Atlantic, specimens have beached on the Argentine coast, the Falkland Islands, Tierra del Fuego, and South Africa. There is a skull in the Te Papa Tongarewa (National Museum of New Zealand) that reportedly came from the Antarctic. (The Dutch stranding [noted above] makes any description of its distribution problematical; it may have been a far-rambling vagrant.) For some reason, New Zealanders like to use the name Scamperdown Whale instead of Gray's Beaked Whale.

These animals are not often observed at sea, but Rice (1978) noted that "the scamperdown . . . has the peculiar habit of sticking its long, needlelike, white snout out of the water as it breaks the surface to breathe." There are many records of strandings, including a mass stranding of 28 animals at New Zealand's Chatham Islands in 1874. From the mass strandings and field observations in New Zealand waters and elsewhere, it can be assumed that this species is more gregarious than other species of *Mesoplodon* and may travel in small groups.

This species is seldom seen at sea, but it is the most common beaked whale to strand in New Zealand (Baker and Van Helden 1999), with 188 stranding events since records began in the 1840s. In June 2001, a mother and calf spent almost five days in Mahurangi Harbour, North Island, New Zealand, giving researchers a rare opportunity to observe *M. grayi* in an unstressed state for more than a passing moment. As the observers (Dalebout, Russell, et al. 2004) wrote, "No obvious confused or stereotypic behavior was observed, as has been the case with many other beaked whales trapped in shallow harbour areas. These situations invariably end in the animals stranding and dying." In this case, the two whales "appeared calm during the majority of the observations, based on their slow, constant swimming speed and regular breathing rate." The mother was estimated to be about 5 m (16 feet) in length, dark grey dorsally, and lighter gray ventrally. "The melon and proximal section of the beak were light gray white. The lower jaw [w]as several centimeters longer than the upper jaw and the protruding portion was white." She had a series of deep indentations along her back behind the dorsal fin, attributed to propeller gashes from a ship strike, appearing to be fully healed. The calf was about 2.5 meters long, half the length of the mother. It was:

> light grey brown dorsally and whiter ventrally. A white patch highlighted the genital region, which was partly visible when the calf performed a half roll at the surface. The tail lacked a centre notch as did that of the adult female. A swath of gray-brown pigmentation continued forward from the blowhole, covering the top of the melon. The sides of the head and the anterior slope of the melon were cream. The eye patches and distal portion of the beak were grey-brown. The beak was relatively short compared with that of the adult female. Overall, this unusual encounter with the cow-calf pair in Mahurangi Harbour, together with at-sea sightings . . . have given us a better understanding of Gray's beaked whales in New Zealand waters.

Accounts of this species must include the single occurrence from the Netherlands. This whale, a 15-foot (4.6 m) female, stranded in 1927, but apparently was not reported in the scientific literature until Boschma described it in 1950. (In 1938, Norman and Fraser wrote that "*M. grayi* is as yet only known from specimens in the southern hemisphere.") Because the location in Holland is over 7,000 miles from the nearest known specimen, Moore (1963) called the event "fantastic" and concluded that "Boschma's specimen . . . must be regarded as a straggler." Other beaked whales have been discovered far from their known ranges, and these discoveries may demonstrate nothing more than our own inabilities to find these elusive creatures at sea.

Hector's Beaked Whale

Mesoplodon hectori (Gray, 1871)

In 1871, J. E. Gray identified the type specimen as representing a new species of *Berardius* and named it *Berardius hectori* after Sir James Hector (1834–1907), an Edinburgh-born geologist and founder of the colonial museum in Wellington. It was subsequently assigned to the genus *Mesoplodon* (Turner 1872; Harmer 1924), where it has remained, in spite of the efforts of McCann (1962b) to reassign it to *Berardius arnuxii.* (He wrote: "*M. hectori* is known only from very young calves [near neonatals] although it was first discovered [1866] nearly a century ago. Conversely, *Berardius arnuxii,* although also known for over a century, is only known from adults!")

In his 1968 classification of the beaked whales, Moore stated that "however sincere McCann may be in his concept of *hectori* representing in fact the young of *Berardius,* he has not succeeded in being objective in his observations . . . and he musters no clear evidence that his concept is correct." Ross's 1970 discussion of the two South African specimens confirmed Moore's conclusion that *hectori* is "a distinct species in the genus *Mesoplodon.*"

In 1938 Norman and Fraser referred to this species as the "New Zealand Beaked Whale," since the only known specimens had been found in that area. The type specimen was from Titai Bay, New Zealand (Gray 1871), and a 9.5-foot animal had been found at Plimmerton, in the same country (McCann 1962b). There was the skull of an immature animal from the Falkland Islands (Fraser 1950), another skull from Adventure Bay, Tasmania (Guiler 1967), and two immature animals from the mouth of the Lottering River, South Africa

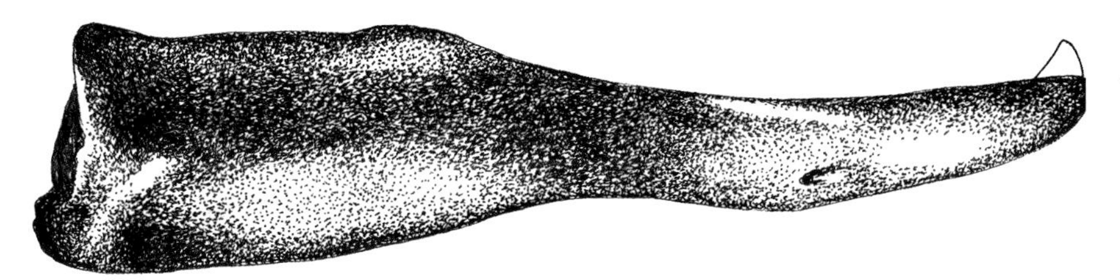

Mandible of *Mesoplodon hectori.*

(Ross 1970). The Tierra del Fuego specimen was discovered in 1975, but its description was not published until 1978 (Goodall 1978). In 1994, a yearling whale, too decomposed to determine the sex, washed ashore in southern Brazil, representing the northernmost record of this species (Zerbini and Secchi 2001). Several other records were reported in Argentina and Chile. *Mesoplodon hectori* used to be one of the least reported beaked whales, but now with the discovery of *M. hotaula* and the resurrection of *M. traversii*, it has lost that distinction.

Hector's beaked whale is still listed among the world's least known cetaceans; as of 2001, Zerbini and Secchi wrote, "published records described less than 30 specimens found ashore." Although *Mesoplodon hectori* is still far from common, enough scattered information has been obtained to enable us to attempt a description and to speculate on its distribution. It is one of the smaller beaked whales, at a body length of about 14 feet (4.5 m) and an estimated weight of 2,000 pounds.

The teeth in adult males are flattened triangles located at the tip of the lower jaw, shaped very much like those of the males of *M. europaeus.* Males are usually scarred by the apical teeth of rival males. On those rare occasions where a fresh carcass has washed ashore, the coloration has been described as dark grayish brown above and pale beneath, with a light blaze behind the eyes and a shortish, light-colored beak. As has been shown with other mesoplodonts, when a live or freshly stranded specimen is described, an often totally unexpected color scheme emerges, so "dark-above-light-below" might be only the beginning (see *Tasmacetus shepherdi*).

Deraniyagala's Beaked Whale

Mesoplodon hotaula Deraniyagala, 1963

In 1963, P. E. P. Deraniyagala described a female beaked whale that had been washed ashore at Ratmalana, Ceylon (Sri Lanka). He named it *Mesoplodon hotaula*. The specific name was derived from the Sinhala words *hota*, meaning "beak," and *ul*, meaning "pointed." He deposited the specimen in the Colombo Museum, of which he was the director. J. C. Moore, while doing a revision of *Mesoplodon*, traveled to the Colombo Museum and determined that the type of *M. hotaula* was in fact a specimen of *M. ginkgodens*. There it sat until the twenty-first century, when a sample of the tissue of an animal from Palmyra Island made its way to M. Dalebout, who was compiling a DNA database on ziphiids. She rapidly realized that it was not *M. ginkgodens* but a new species. She and a group of cetologists began describing it. Charles Anderson, one of the members of the group, traveled to Sri Lanka and obtained a DNA sample from Deraniyagala's specimen. It turned out to be identical to the individual from Palmyra, and both are now identified as *Mesoplodon hotaula*.

Deraniyagala's beaked whale is now known from eight individuals from the tropical North Pacific and Indian Oceans. Adult males are darkly pigmented over the entire body, with the exception of a lighter area on the ventral surface of the head. The one adult male that was seen freshly dead had a small area of white pigmentation just posterior to the tooth and did not have white parasite scars. Adult males appear to lack the linear scars typically seen in males of other *Mesoplodon* species. Females are probably medium gray dorsally and light gray ventrally. The moderate-length beak slopes fairly

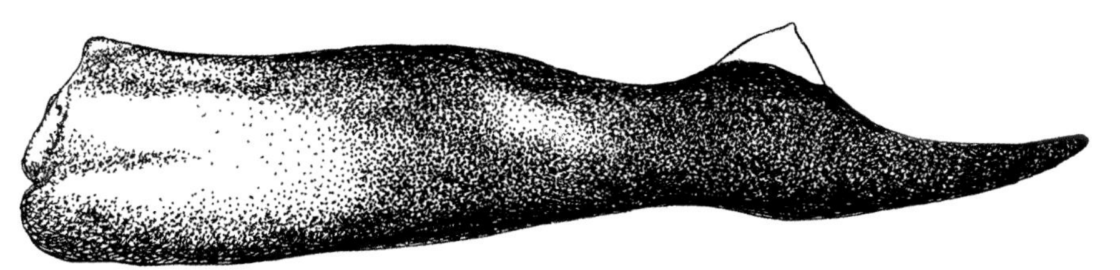

Mandible of *Mesoplodon hotaula*.

steeply onto a small though well-rounded melon. In males, the mouth line curves sharply upward about halfway back from the beak tip. This arch is less pronounced in females and juveniles, although the mouth line is still sinusoidal. The adult males have a small portion of the tip of a tooth exposed just anterior to the apex of the mouth line. Adult body length ranges from 3.8 to 4.8 meters. Recorded maximum body length for adult males and females is 4.3 meters and 4.8 meters, respectively. Length at birth is unknown.

A single pair of teeth is situated about 3 cm posterior to the symphysis. The teeth are triangular, with a convex lower border. In adult males the teeth are about 10 cm long and 9 cm wide. In females the teeth are about 5 cm long and 5 cm wide. Only the tip of the tooth is exposed in adult males. In the one adult male that was observed fresh, the tip of the tooth erupted about 2 cm. In females and juveniles the teeth are not exposed. The teeth resemble the leaf of a ginkgo tree. The denticle is centrally positioned between anterior- and posterior-most points of the tooth. Anterior and posterior tooth margins are slightly concave in females and nearly symmetric. In males, the anterior and posterior margins are convex. The only two males that are known have the tip broken and barnacles on the tip. The alveolus of the teeth on adult males does not overlap the symphysis, which separates this species from *M. ginkgodens.*

Baumann-Pickering et al. (2010) recorded echolocation signals at Palmyra Atoll from an unknown mesoplodont that they suggested might represent *M. hotaula*. It seems as though they were correct.

Strap-Toothed Whale

Mesoplodon layardii (Gray, 1865)

In a 1988 review of this species, Goodall, Folger, and Lichter wrote, "The Layard's beaked whale [strap-toothed whale] is known almost exclusively from strandings." There are 190 records of this species in the Smithsonian Cetacean Distributional Database, more than any other beaked whale. People may be more inclined to report these strandings because of the unusual nature of the teeth. We have 56 records from Australia, 55 from New Zealand, 6 from Chile, 3 from Brazil, 2 from Uruguay, 11 from Argentina, 14 from the Falkland Islands, 2 from South Georgia, 2 from Namibia, 23 from South Africa, and 1 from Myanmar.

The Uruguayan specimen, a 19-foot-long female, was found near La Paloma, and is the northernmost record for this species (Praderi 1972). Very little is known about the life of this creature, but one thing is obvious: it is exclusively a Southern Hemisphere species. As with *M. europaeus* (see p. xx), if a species strands often enough, every event will not be recorded in the scientific literature. Mead (1989c) listed 27 previously unpublished strandings of *M. layardii* from Australia, South Africa, the Falkland Islands, and New Zealand.

One of the largest of the mesoplodonts, *M. layardii* has been recorded to reach a length of over 19 feet (Waite 1922), but most of the other known specimens have been between 13 and 17 feet long (Hale 1939). The longest record that we have is 6.25 meters (20.5 ft); it was a female that stranded on the first of April, 1976, in Auckland, New Zealand, and was reported by Anton Van Helden.

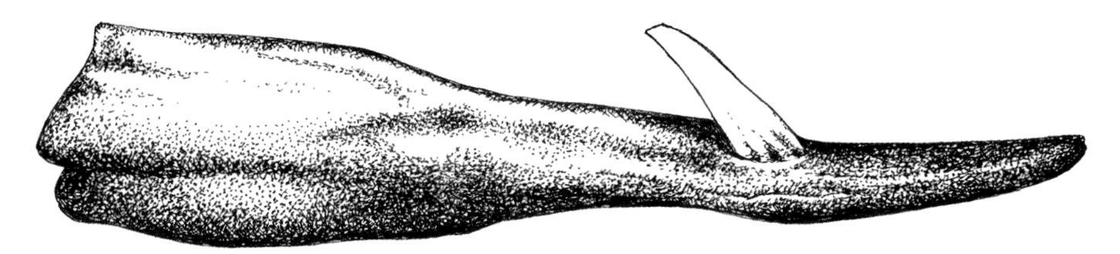

Mandible of *Mesoplodon layardii.*

The color of a living animal has been described as follows: "bronze brown on the dorsum, pale gray ventro-laterally grading to off-white on the underside" (Hale 1939). Forty-eight hours after this stranded animal died, it became "black above, grading to purplish pink with isolated gray patches on the side of the head." The difference in these two descriptions demonstrates the postmortem changes that occur in the external appearance of whales and may very well account for the predominance of descriptions of stranded whales as "black," since in most species the skin color darkens appreciably with even short-term exposure to the sun.

In life, the strap-toothed whale is easily the most dramatically colored of all the mesoplodonts and ranked with the killer whale as the most elegantly attired of all cetaceans. Baker describes the color scheme as follows: "Generally dark blue-black on the back and flanks, with a white beak and throat, and a white patch around the genital area. In adult animals there is a greyish 'cape' over the back between the head and the dorsal fin. The white throat patch has a median ventral point, directed tailwards between the flippers. The genital patch has two forward pointing extensions on either side of the ventral midline. Another unusual feature is a small, pointed flag-like pale spot on the flanks just underneath the flipper."

Young animals are not easily distinguished from many other species of *Mesoplodon*, since they have the regular characteristics of the genus, including paired throat grooves converging toward the beak, a small dorsal fin and flippers, and flukes that are not notched. In the immature specimens the lower teeth, located toward the posterior border of the jaw, are triangular and flattened, much like those of any other *Mesoplodon*. With maturity, however, adult males of this species develop teeth that make them unique among whales and probably in the entire animal kingdom. The small triangular tooth begins to curve outward at first, "giving the appearance of the teeth having the concave surface outwards" (Oliver 1924). From this stage, where the denticles are pointing away from the lower jaw, they then curve up, back, and over the upper jaw, until they form an arch over the rostrum, finally preventing the animal from opening its mouth to any significant degree. This arrangement is so bizarre that Goodall et al. (1988) wrote, "perhaps we are seeing only animals that died because they could not open their mouths (it has been stated that these animals were in good health, but few have been fresh and no stomach contents examined for food remains)." Even though Heyning and Mead's revelations about suction feeding would not be published for eight years, Goodall and colleagues guessed that "perhaps it is not so difficult to slurp squid or whatever with a restricted mouth."

These extraordinary teeth have been likened to "the tusks of a boar" (Gaskin 1968), and Moore (1968) wrote that "the teeth of *Mesoplodon* attain great length in the second largest species of the genus, *M. layardi* (usually written *M. layardii* by most authors). The longest personally measured is 330 mm (12.9 inches)." (By comparison, the "tusks" of the species *M. densirostris, M. stejnegeri,* and *M. carlhubbsi,* known as the saber-toothed whales, rarely exceed 6 inches [150 mm] in height.) From the examination of the stomach contents of stranded specimens, it has been determined that strap-toothed whales eat squid, and because they are suction feeders like all other mesoplodonts, the males must suck in the cephalopods through the reduced opening of the mouth—rather like eating squid through a straw. Juveniles and adult females have no such problem, since their teeth do not develop in this amazing fashion.

The food habits of strap-toothed whales were examined in detail using stomach contents from 14 stranded whales found on South African and New Zealand coasts. Although a few unidentified fish otoliths and crustacean remains were found in two of these stomachs, 24 species of oceanic squids (some of which occur at a great depth) accounted for 94.8% of counted prey items. Prey sizes were compared between males with fully grown strap-teeth and females/immature males without erupted teeth, using dorsal mantle lengths and weights of squids estimated from beak measurements. Although females/immature males ate longer squids than males, there was no significant difference in the estimated weights of squids eaten by the two groups. The presence of fully erupted teeth in adult males, therefore, did not seem to influence the size of prey ingested, even though an adult male could only open its jaws about half as wide as a female (Sekiguchi, Klages, and Best 1996).

Although McCann (1964) published a description of the female reproductive organs of this species, in which he wrote that "the calving of *M. layardi* takes place in the early spring of the Southern Hemisphere," almost nothing else is known of the life cycle of this species. There is one recorded observation of this species at sea: Gaskin (1968) wrote that "one male was recognized by the author. . . . the tooth on the left side of the head was clearly seen. This animal was one of a group of three, of which all were presumed to be of this species." Observers at Victor Harbor, Australia, saw two animals "sporting offshore" shortly before a male and female *M. layardii* beached themselves. The male was 15 feet long and the female 2 feet longer. Hale (1931a), who recorded the observations, indicated that the animals might have been mating.

North Atlantic True's beaked whale

Southern Hemisphere True's beaked whale

True's Beaked Whale

Mesoplodon mirus True, 1913

In his original description of this species in 1913, F. W. True described this color pattern: "back slate black, lower side yellow purple flecked with black, median line of the belly somewhat darker; a grayish area in front of the vent; fins the color of the back." Subsequent authors have been less inclined toward "yellow purple," giving instead the ventral coloring as "light gray" or "slate-colored." Fraser (1946) quoted a description of a 15-foot-10-inch male that was "purplish white in color," lightest on the dorsal surface. When cetaceans are exposed to sunlight, the color often darkens quickly, so frequent reports of black-bodied animals may not be accurate descriptions of their color in life.

Mesoplodon mirus (the Latin name *mirus* means "wonderful" or "amazing") is not a particularly large species, reaching a known maximum length of 17.5 feet (5.3 m) (McCann and Talbot 1963). The two teeth of the male are located at the very tip of the mandible and angle slightly forward. They are heavier and thicker than those of the female, whose teeth are buried in the gums. The position of these teeth in the male led McCann (1974) to suggest that "this species alone of the *Mesoplodon* would inflict parallel scars on its opponent." The other species, with their "defensive teeth" located in a less convenient position, would inflict "single linear scars at a time on their opponents." Even in males, however, the teeth are relatively small for the genus, rarely reaching a total length of 2 inches (50 mm). True's beaked whale has a proportionally small head with a pronounced forehead bulge and sharply

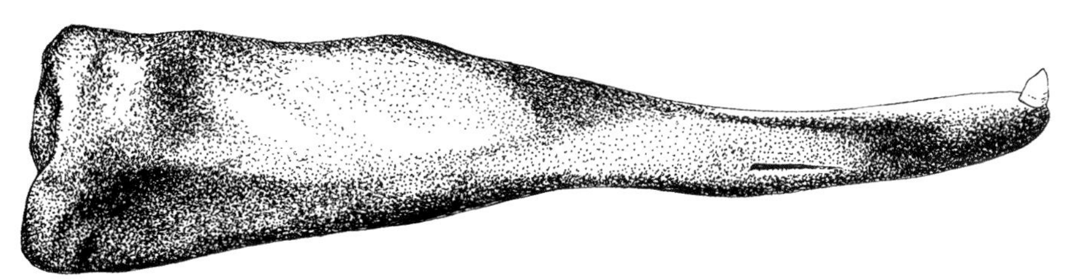

Mandible of *Mesoplodon mirus*.

defined throat grooves. In most specimens there is no notch in the tail fin, but Nishiwaki (1972) claimed that "the tall fluke is wide with a small notch at the posterior edge." The presence of a notch in the flukes is variable, and therefore it is not a reliable character (Ryder 1887).

There are two species of *Mesoplodon* in the North Atlantic, *M. mirus* and *M. europaeus,* that have two teeth at or near the front of the lower jaw. In analyzing their differences and similarities, Moore and Wood (1957) wrote that there is "evidence of some segregation between the species, although there is also considerable overlap." The distributional differences seem to lie at the extremes of the range in the western North Atlantic. *M. mirus* has Cape Breton Island, Nova Scotia, as its northernmost limit, whereas *M. europaeus* has been found only as far north as Long Island, New York. Flagler Beach, Florida, is as far south as *M. mirus* seems to go, but *M. europaeus* is known from Trinidad (Fraser 1955) and Jamaica (Rankin 1953).

These scattered strandings would appear to suggest a solitary existence for the species, but the small number makes the sample less than useful. (In a summary of the strandings, Moore [1966] listed a total of twelve—eight on the coast of North America and four in British waters.) A 17-foot female that beached on the North Carolina coast carried a 7-foot 2-inch fetus, apparently full term (Brimley 1943). With this exception, we know hardly anything about the life history of the species. One must remember that strandings are unusual occurrences, and they do not represent the normal behavior of an animal, nor do they necessarily define its range. Those animals that do come ashore are probably sick or stressed, and we therefore must not interpret this behavior as representative or diagnostic of the species.

True's beaked whale has been recorded from the temperate waters of the North Atlantic, from Nova Scotia to Florida in the west, and from the Outer Hebrides to the west coast of Ireland in the eastern North Atlantic. When a 15-foot female stranded in November 2009, on Buckroe Beach in Hampton, Virginia, a biology class from the local high school was brought out for the once-in-a-lifetime opportunity to see one of the world's rarest animals.

In addition to the well-documented North Atlantic population, there is also a Southern Hemisphere population, which may constitute a subspecies. The species appears to be centered in the North Atlantic, where there are twenty-eight strandings from the United States, one from Nova Scotia, one from the Bahamas, six from Ireland, one from France, and one from Portugal. In the North Pacific, there are questionable reports from Guam and the Marshall Islands. The population in the Southern Hemisphere appears to be centered in the Indian Ocean, where there are ten from Australia, eight from South Africa, and one from Mozambique. In the South Atlantic there is one from Brazil, and in the South Pacific Ocean, one from New Zealand.

Perrin's Beaked Whale

Mesoplodon perrini

Dalebout, Mead, Baker, Baker, and Van Helden, 2002

This is one of the newest species of beaked whales to be described or resurrected. The first two specimens were found in May 1975 stranded on the California coast, with two more specimens found in 1978 and 1979 and the last in September 1997. They were initially identified as Hector's beaked whales (*M. hectori*)—the only records of *M. hectori* from the Northern Hemisphere. The 1997 specimen, a 6-foot juvenile, was found stranded and run over by an amphibious tracked vehicle on the beach of the Camp Pendleton Marine base near Monterey, California. An mtDNA sequence database of beaked whales later showed that all the California specimens were distinct from *M. hectori* (Dalebout et al. 1998) and belonged to an undescribed species. The new species turned out to be genetically related to the pygmy beaked whale (*M. peruvianus*), the next most recently described species, and probably represents its Northern Hemisphere sister species. Then again, maybe not. *M. peruvianus* has now been recorded from Mexico, so the antitropical species pair idea is not valid (Urbán-Ramirez and Aurioles-Gamboa 1992).

The male California specimens and the Southern Hemisphere *M. hectori* have similar teeth—two large triangular teeth at the apex of the lower jaw—but the actual shape of the jaw is somewhat different. As with many mesoplodonts, confirmation of identity of any specimens is dependent on subtle skull characteristics and genetic markers. To date all the known specimens have been found on the coast of California, implying that its core range is focused in that area. Although there are no confirmed live sightings, some

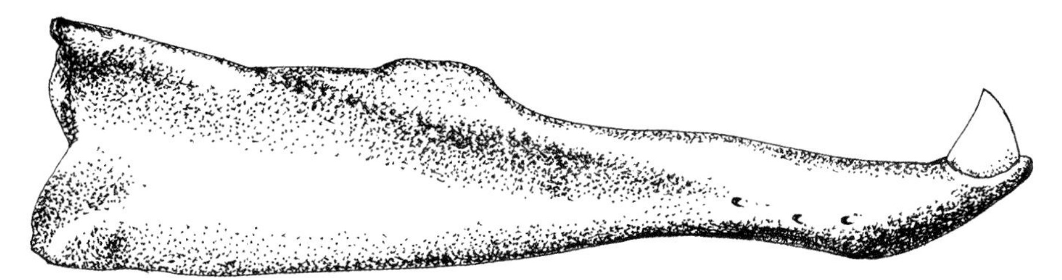

Mandible of *Mesoplodon perrini*.

sightings of unidentified mesoplodonts off the central California coast have been tentatively attributed to this species. Dalebout et al. (2002) named the new species *Mesoplodon perrini* after American cetologist William Perrin, who had retrieved the first Camp Pendleton specimen, and six days later, from the same locality, an adult female thought possibly to be the mother.

In the Beaked Whale issue of the American Cetacean Society's *Whalewatcher*, Bill Perrin recounted the story of "The Accidental Whale," the discovery of the beaked whale that would be named for him. A beaked whale on the beach is always worth an investigation, so when a marine reported the squashed neonate on the beach at Camp Pendleton, Perrin and several colleagues rushed to the site from the National Marine Fisheries Service lab in La Jolla. When the presumed mother also stranded, Perrin collected those remains as well and had both specimens shipped to the Smithsonian. Working with skeletal remains and mtDNA, researchers from the Smithsonian and New Zealand sequenced the mtDNA of the various specimens (including those that had been previously identified as *M. hectori*) and established the new species, *Mesoplodon perrini*. When the specimens were put on exhibit at the Los Angeles County Museum, John Heyning, then the museum's deputy director, said, "It's clear that even for whales, those large animals that everyone loves, there's a lot we don't know." Perrin wrote, "That about sums things up; if you come across a piece of carrion on the beach, pick it up and take it home. It may be a new species (and it might even be named after you)."

We now have six records of Perrin's beaked whale, all from California. An adult female measured 14.5 feet (4.4 m) and an adult male 12.8 feet (3.9 m). Baumann-Pickering et al. (2014) recorded signals from a beaked whale off California that they surmised might have been *M. perrini*.

Pygmy Beaked Whale

Mesoplodon peruvianus
Reyes, Mead, and Van Waerebeek, 1991

The pygmy beaked whale is one of 4 species that have been discovered or resurrected in the past 15 years (the others are *Mesoplodon perrini, Mesoplodon hotaula,* and *Mesoplodon traversii*). It is the smallest known beaked whale, with adults being only about 4 meters (13 ft) long. There may be differentiation in the Northern and Southern Hemisphere populations. The largest specimen in the Southern Hemisphere was a 372-cm (12.2 ft) male and in the northern hemisphere a 407-cm (13.4 ft) specimen that was measured photogrametrically by Pitman and Lynn (2001). It also has the ignominious distinction of being the only marine mammal described as a consequence of fisheries bycatch: 7 of the first 10 specimens described were caught in the shark drift-net fishery off the coast of Peru. Three were described from stranded specimens. According to the original description by Reyes et al. in 1991, from the skull characteristics, "the species that *M. peruvianus* most closely resembles are *M. hectori* and *M. stejnegeri*." The first ten specimens of *M. peruvianus* were collected from the coast of Peru, but the species is not restricted to Peruvian waters.

The body of the pygmy beaked whale is typically spindle shaped, with an unusually thick tail. The melon is somewhat bulbous and slopes down into a rather short beak. The mouthline in males has a very distinct arch, with two teeth protruding slightly from the gumline before the apex. The coloration is typically dark gray above and lighter below, with little scarring. The dorsal

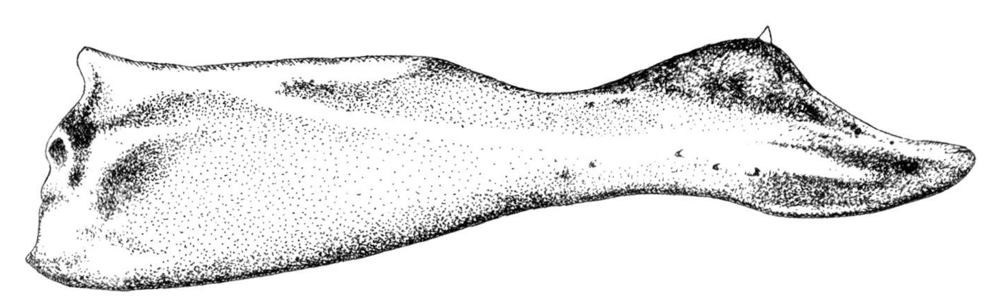

Mandible of *Mesoplodon peruvianus.*

fin is not falcate like those of other mesoplodonts, but small, triangular, and wide-based, like that of the harbor porpoise. The teeth are extremely small and egg shaped in cross section, generally not visible in sightings at sea. Males can be distinguished by a contrasting cape pattern on the back.

In 1987, Pitman, Aguayo, and Urbán published "Observations of an Unidentified Beaked Whale in the Eastern Tropical Pacific," in which they described multiple sightings of mesoplodont whales, which were seen to have two distinct color morphs: "a conspicuously marked black and white form, and a relatively nondescript morph that appears to be uniformly gray-brown. When they have been seen together, the black and white animal has been judged by several observers to be the larger of the two." The larger animal, presumed to be a male, had a broad cape of white or cream color aft of the flippers and swooping down before the dorsal fin. Close observation revealed multiple scratches and scars on the larger animal, confirming that it was indeed a male. As with most female mesoplodonts, the smaller, nondescript animals had no scarring. From the at-sea sightings this is known to be a relatively distinctive short-beaked species.

This species is believed to be the same as *Mesoplodon 'A,'* which was identified from a series of sightings during surveys of the Eastern Pacific in the 1980s and 90s, which at the time did not conform to the appearance of any known species. In 1992, Urbán-Ramirez and Aurioles-Gamboa published a record of two carcasses that had been found in Bahía de la Paz, Baja California, in the spring of 1990. (Bahía de la Paz is on the east coast of the Baja Peninsula, on the Gulf of California side.) Both were about 3.4 meters long, and skeletal examinations showed them to be adults. Adult beaked whales 13 feet long are very likely to be *M. peruvianus.*

By 2001, some 65 at-sea sightings had been made in the Eastern Tropical Pacific warm waters from Mexico to Peru, suggesting it to be largely restricted to this area; however, a stranding has since been documented from North Island, New Zealand (Baker and van Helden 1999), suggesting a much wider distribution in the tropical Pacific. Currently we have 33 records of this species in the Northern Hemisphere along the coasts of North America and 17 records in the Southern Hemisphere, 15 from Peru and 1 from New Zealand. As better records accumulate, we would not be surprised to see this species in the western half of the North Pacific. It appears to be restricted to the Pacific.

Stejneger's Beaked Whale

Mesoplodon stejnegeri True, 1885

Stejneger's (pronounced "*sty*-ne-gers") beaked whale is one of the little-known saber-toothed beaked whales from the North Pacific Ocean. In 1938 there were "but two specimens, both from the Pacific coast of North America" (Norman and Fraser 1938). One of these was the type specimen, described from a skull found by Leonhard Stejneger on Bering Island in 1883, and the other was the first full specimen to be observed, a 17-foot male that washed ashore at Yaquina Bay, Oregon, in 1904 (True 1910). The Yaquina Bay specimen was described thus in a local newspaper (quoted in True): "On either side of the mouth are two villainous-looking tusks, several inches in length. They are at the back of the mouth, and extend up to a level with the top of the jaw. They are very wide and flat, squared on top . . . The head is equipped with a blowhole, like that of a whale. The eyes are very low, almost underneath the lower jaw." One of these tusks "measured 5 ¾ inches along its anterior border, 8 ¼ inches along its posterior border, 3 ¼ inches from anterior to posterior border, and just over ½ inch in tra[ns]verse thickness" (Norman and Fraser 1938).

We now have 146 confirmed records and 6 probable records. Restricted to the North Pacific Ocean, this whale has been recorded from the Commander and Pribilof Islands in the Bering Sea, south to the Sea of Japan in the west, and to San Diego, California, in the east. From the stranding records it appears that the range of Stejneger's beaked whale is more northerly than

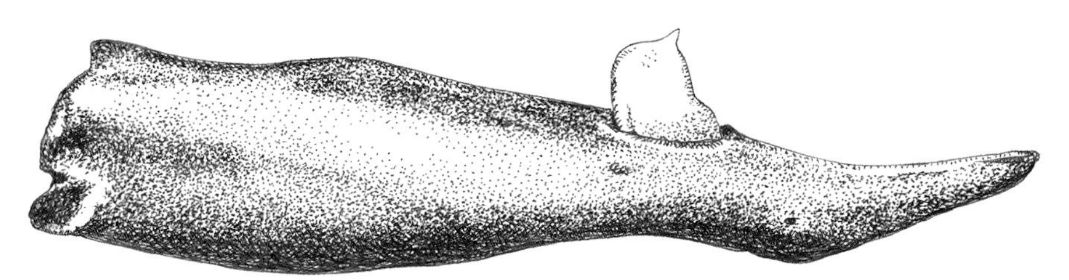

Mandible of *Mesoplodon stejnegeri.*

that of Hubbs' beaked whale, though they overlap in areas of the western United States and northern Japan.

Males of Stejneger's grow up to 19.7 feet (6.0 m) and females 17.8 feet (5.4 m) (Smithsonian Cetacean Distributional Database).

The species is usually dark gray with scars and whitish patches, and the jaws anterior to the teeth may be white. Two specimens were described and illustrated by Nishimura and Nishiwaki (1964). The 17-foot (5.2 m) male was "dark gray all over" with white scars, and the 7.8-foot (2.4 m) female was "grey on the dorsal surface as well as the areas around eyes and snout, and the upper surfaces of flippers and tall flukes, but whitish on the ventral side of the body including the under surface of flippers and tail flukes." From the illustrations it can be seen that the scars on the male's body are mostly paired parallel lines, and the authors made the "supposition that such streaks are nothing but scars formed by other full-grown males during plays or fightings. The female specimen, on the other hand, showed no traces of such streaks." Unlike the other "saber-toothed" whales (*M. bowdoini* and *M. carlhubbsi*), the lower teeth of *M. stejnegeri* are fully exposed. (In females of all three species the lower teeth do not erupt at all, so the shape of the tooth and the skull proportions must be the differentiating factors.)

From 1975 to 1994, twenty-three Stejneger's beaked whales stranded at Kuluk Bay, Adak Island, in the Aleutians. Four of these were adult females, found in late August 1994, on a 1-mile stretch of beach adjacent to Clam Lagoon, and described in detail in 1999 by William Walker and Bradley Hanson in *Marine Mammal Science*. Along with the females, three unborn fetuses were also recovered, all of which provided a heretofore unavailable glimpse at some of the details of the lives (and deaths) of *M. stejnegeri*. All the adults were dark brownish-gray on the back and sides, fading to a lighter underside. The undersides were heavily mottled and marked with numerous scars of cookiecutter shark bites, but perhaps the most startling element in the color scheme of these whales was the underside of the flukes, marked with "white

An adult male Stejneger's beaked whale that stranded on Graham Islands, British Columbia, showing the extensive scarring. Photo taken by Charlotte Tarver, courtesy of John Ford.

Ventral side of the flukes of a Stejneger's beaked whale that stranded at Kuluk Bay in 1994.

concentric striations seemingly coalesced into a symmetrical, solid white anchor-shaped blaze." The "blaze" resembles a roman candle or some kind of umbrella-like firework lighting up the black sky with cascading white lights. This pattern is also seen in the type of *M. carlhubbsi,* adult male (USNM 278031) and adult female (USNM 504128). It was seen in a recent specimen of *M. perrini* and resulted in its misidentification. Rarely mentioned in descriptions of beaked whales are the "flipper pockets," indentations where the flippers can be tucked back along the sides, reducing drag as the animal dives. On *M. stejnegeri,* these pockets are darkly pigmented, which makes them look like flipper *shadows.*

In their 2013 discussion of the strandings in the Aleutians, Baumann-Pickering, McDonald, et al. wrote that "Since the first record of a mass stranding in 1975, a mass stranding of this species with two or more animals has occurred on an average of every 3 years in the Aleutians. This is an alarmingly high number of mass strandings (11 with two or more animals) from unknown causes with potential population risks especially considering the strong likelihood that some mass strandings on these remote islands are missed." To date, whatever we know about Stejneger's beaked whales we have learned from dead animals, so to obtain an idea of what the beaked whales might be doing in the area, the authors deployed a bottom-moored recording system in the Sitkin Canyon (in the Aleutian Trench, off the Rat Islands) from June 3 to August 26, 2010. They wrote, "Due to the remote locations and inaccessible sea during much of the year, long-term autonomous acoustic monitoring may lead to further understanding of the behavior and ecology of this understudied species of beaked whale, and might provide guidance for mitigating impacts leading to mass strandings." Presumed sounds from Stejneger's beaked whales were obtained from these recordings.

Spade-Toothed Whale

Mesoplodon traversii (Gray, 1874)

In 1874, J. E. Gray described a new species of beaked whale from the Chatham Islands as *Mesoplodon traversii*, based on a jaw that Sir James Hector (1873) thought was a specimen of *M. layardii*. More than a century later, Reyes et al. (1995) described a beaked whale skull from the Juan Fernández Islands, 375 miles off the coast of Chile, as a new species, *M. bahamondi*. In 1999, Alan Baker and Anton van Helden described another *Mesoplodon* skull, this one found on White Island (New Zealand), as *Mesoplodon ginkgodens*. In 2002, however, after analysis of the DNA of the White Island specimen, the Juan Fernández specimen, and the jaw bone of *M. traversii*, it was realized that they were all of the same species, Gray's *M. traversii*. The skulls from Juan Fernández and White Island were the first ever recognized as *traversii*. So, once there was a beaked whale known as *Mesoplodon bahamondi*, but now the name resides only in the convoluted nomenclatural history of the spade-toothed whale. In summing up their findings, Thompson et al. (2012) wrote, "Our results combine morphological and molecular evidence to unify three fragmentary and disparate museum-held beaked whale specimens resulting in the synonymy of *M. bahamondi* with *M. traversii*. This multidisciplinary approach has provided new insight into the distribution and distinctiveness of what is arguably the least known of all living species of cetaceans."

In December 2010, a mother and calf *Mesoplodon* were found alive on Opape Beach (North Island), New Zealand, but they subsequently died. Scientists took photographs and tissue samples and thought that the pair might be Gray's beaked whales (*M. grayi*), which is the species that strands most

Mandible of *Mesoplodon traversii*.

frequently on New Zealand's shores, but the Opape Beach whales correlated genetically with *Mesoplodon traversii*. As Gray (1874) did not give the specimen a common name, van Helden et al. called it the "spade-toothed whale" because, they wrote, "the extraordinary shape of the tooth . . . is reminiscent of the oblong blade of the flensing knife used by American whalers in the 19th Century." This most recent stranding is the first time the species has ever been known in the flesh. The news of the discovery was broadcast all over the world, usually with the headline "The World's Rarest Whale." That was also the title of the 2012 article in the journal *Current Biology* by Thompson, Baker, van Helden, Patel, Millar, and Constantine, where the two specimens were described and their DNA analyzed. They wrote: "Two individuals of this species, previously known only from two skull fragments and a mandible, were recently discovered beach-cast in New Zealand. Although initially mis-identified, we have used DNA analysis to reveal their true identity. We provide the first morphological description and images of this enigmatic species. This study highlights the importance of DNA typing and reference collections for the identification of rare species."

In addition to the DNA differentiation, the authors noted that *Mesoplodon traversii* can be differentiated from the very similar *M. grayi* (the original misidentification) by a more pronounced melon, a dark beak, a dark eye patch, white belly, and dark flippers. They concluded, "We can now confirm that the spade-toothed whale is extant and for the first time we have a description of the world's rarest and perhaps most enigmatic marine mammal."

Shepherd's Beaked Whale

Tasmacetus shepherdi Oliver, 1937

In 1933, George Shepherd, the curator of the Wanganui Museum, read in a local newspaper about a whale washed up on the beach near Ōhawe on the south Taranaki coast of New Zealand's North Island. Shepherd visited the site, collected the skeleton, and became convinced that it was a new species of beaked whale. The director of the Dominion Museum, W. R. B. Oliver, confirmed his conjecture and named the species after Shepherd. Oliver called it *Tasmacetus,* not for Tasmania, which is more than 1,000 miles from New Zealand, but for Abel Tasman (1603–59), a Dutch explorer best known for his voyages in the service of the Dutch East India Company. He was the first European explorer to reach the islands of Tasmania and New Zealand. The Tasman Sea separates New Zealand from eastern Australia.

Even in the realm of beaked-whale dental peculiarities, *Tasmacetus shepherdi* stands out. It is the only beaked whale with a full set of erupted teeth in addition to the large pair at the apex of the lower jaw. The apical teeth affiliate this species with the beaked whales, but the additional teeth cause it to be placed in a separate genus. At least five species of beaked whales are occasionally found with a mouthful of small teeth, but they are below the gumline and are not considered functional (Boschma 1951a). Male and female *Tasmacetus* have teeth in both jaws: seventeen to twenty-one pairs in the upper jaw and eighteen to twenty-seven pairs in the lower, including the apical teeth, which are erupted only in males. Like all other beaked whales, *Tasmacetus* has paired throat-grooves that facilitate suction feeding, but it is not clear how the teeth might be employed in this method.

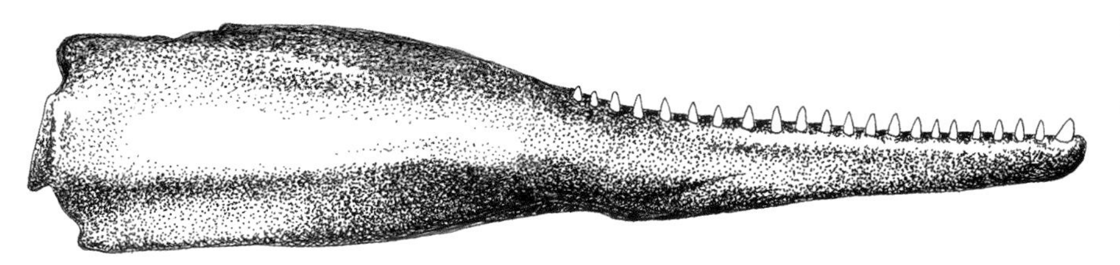

Mandible of *Tasmacetus shepherdi.*

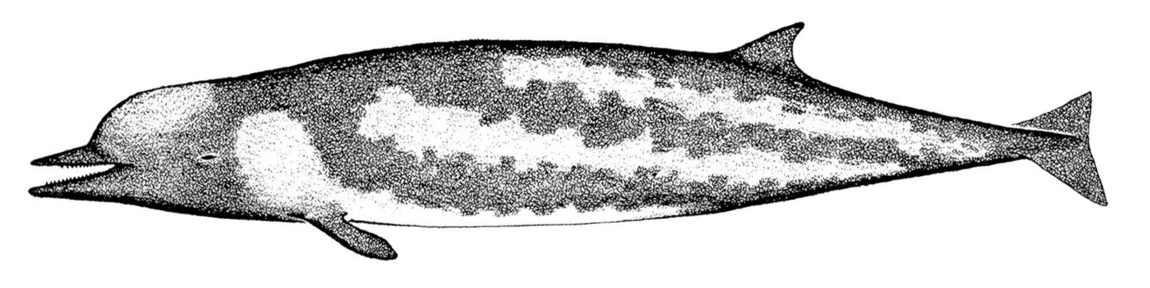

Shepherd's beaked whale

We now have 36 confirmed records of Shepherd's beaked whale. Males and females reach up to 21.7 feet (6.6 m). There is one record of an adult male at 30 feet (9.1 m) long (Sorenson 1940), but all the other specimens were considerably smaller. The forehead appears to be more pronounced and the rostrum more pointed than in the various species of *Mesoplodon* (Mead and Payne 1975). "It is unique among beaked whales so far examined," wrote Pitman et al. in 2006, "in that males and females share the same well-marked color pattern."

While sitting on a bluff overlooking the sea in Sumner, a suburb of Christchurch, New Zealand, W. A. Watkins (1976) made what was considered the first observation of this species in the wild by a cetologist. His field notes, quoted in his paper describing this sighting, read as follows: "wavy, light and dark streaks (2 each?) from shoulder down below fin along sides, beaked, light on top of head, dark back and triangular fin, short flippers, light belly."

Until a specimen was discovered washed ashore at Peninsula Valdés, Chubut, Argentina, it had been assumed that this species was restricted to New Zealand waters. As of 1972, all five known specimens had been recorded from those islands: two stranded at Mason's Bay, Stewart Island, in February 1933, one (the type specimen) at Wanganui in October 1933; and two more from the Avon-Heathcote estuary and Birdling's Flat in March 1951 and in 1962, respectively (Gaskin 1968). The discovery of a female in Argentina by Mead and Payne "extended the range to another ocean," and in 1976, Brownell and colleagues published a report documenting the species in the eastern South Pacific, thousands of miles from its supposed restricted range. A cranium, mandible, and cervical vertebra were found on the beach at Loberia Vieja, in the Juan Fernández Islands, off the coast of Chile. This discovery "now suggests that *T. shepherdi* may have a circumpolar distribution in the southern temperate oceans, as do many other species of cetaceans." This suggestion was verified by the seven individuals of this species that stranded on Argentine coasts in recent years. The earliest one was found at Peninsula

Valdés by Mead and Payne in 1973; the most recent one stranded in Santa Cruz Province in September 2003 (Grandi et al. 2005).

People who study beaked whales sometimes get really lucky and see them at sea. Aboard the U.S. Coast Guard icebreaker *Polar Star* en route to McMurdo Sound in the Antarctic in December 2005, Bob Pitman (and his shipmates on deck)

> saw a group of suspiciously small whales rolling at the surface about a kilometer ahead . . . as they all arched up fairly high for their final dive, I got a clear view of one animal's beak and an unusual color patterning. *Tasmacetus*! After my 30 years at sea and 13 trips to Antarctic, they finally stepped out from behind the curtain . . . The only previously confirmed live sighting of Shepherd's beaked whale was in 2002 at Gough Island in the South Atlantic. Like many other species of beaked whales, it isn't known if Shepherd's beaked whales are really rare or just rarely identified. They inhabit some of the stormiest oceans in the world, so we may never know the answer to that question.

In 2006, with van Helden, Best, and Pym, Pitman published an article on the appearance and biology of this rarely seen species, based on strandings and at-sea observations. *Tasmacetus* is so rare that every confirmed observation can be listed. Pitman et al. discounted Watkins' 1976 sighting by saying "the description provided does not rule out any of several beaked whale species known from the region." Numbers 1 and 2 were photographed by South African cetologist Peter Best from a helicopter over Tristan da Cunha in November 1985. Sighting number 3 was made by naturalist Antony Pym at Gough Island in 2002. Number 4 was Bob Pitman's shipboard sighting noted above.

In the waters of South Australia, about 70 miles southwest of Adelaide, Kangaroo Island is Australia's third-largest island, after Tasmania and Melville Island. During a June 2013 survey of cetaceans off the west coast of the island, International Fund for Animal Welfare researchers spotted (and photographed) a small group of *Tasmacetus shepherdi*, which, from the photographs, appear to be dark slate gray, with a lighter area aft of the dorsal fin. With variations having to do with gender, age, and viewing conditions, all described *T. shepherdi* as a slim, long-beaked whale with a prominent forehead, identifiable at sea by a pale shoulder patch, conspicuous pale melon, and long dark beak. If you should happen to be flying over a school of Shepherd's beaked whales, you will know them by "the white melon, black back extending from about the blowhole to the mid-dorsal fin region, and a paler gray back posterior to the dorsal fin."

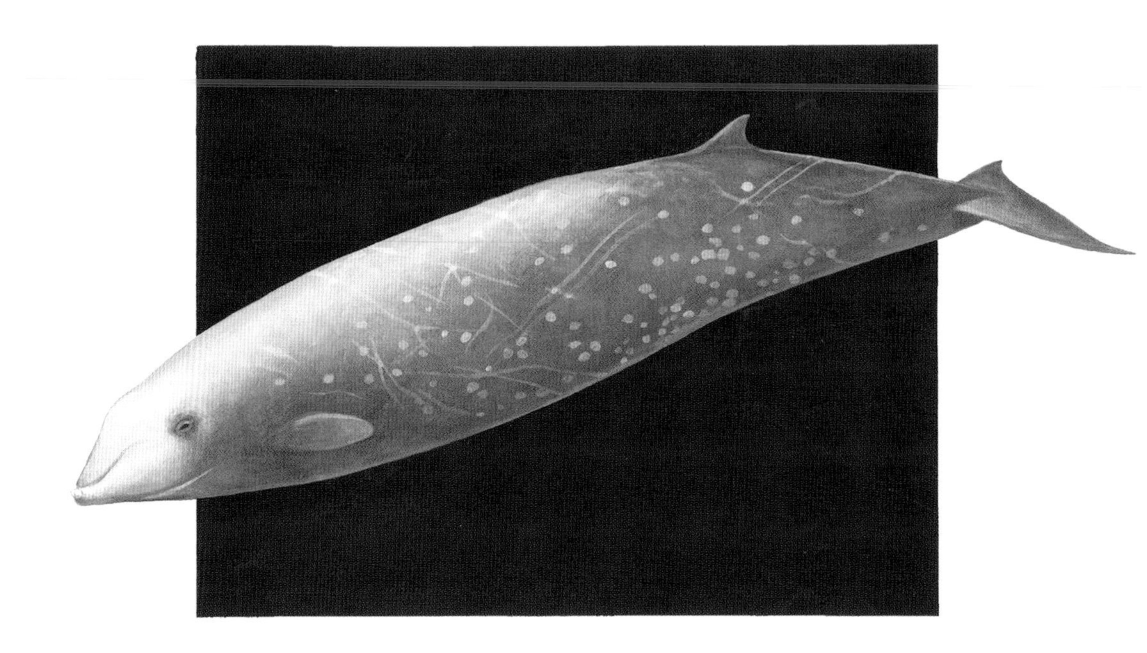

Cuvier's Beaked Whale (Goosebeak Whale)

Ziphius cavirostris G. Cuvier, 1823

When a whale skull was unearthed in Fos (Provence) in 1803, it was brought to Georges Cuvier (1769–1832), France's foremost zoologist, and the founder of comparative anatomy. The bones of the skull were so densely ossified that Cuvier believed it was petrified and therefore a fossil (Rice 1998). But in an 1872 article entitled "On the Occurrence of *Ziphius cavirostris* in the Shetland Seas, and a Comparison of Its Skull with That of Sowerby's Whale (*Mesoplodon sowerbyi*)," Sir William Turner, a British anatomist, wrote,

> The illustrious Cuvier, in his treatise "*Sur les Ossemens fossiles*," described and figured an imperfect skull which had been obtained in 1804, by M. Raymond Gorsse in the department of Bouches-du-Rhône, near Fos, on the southern coast of France . . . Cuvier recognized it to belong to an undescribed genus of cetaceans, to which he gave the name of *Ziphius*; and from the deep hollow which it possessed at the base of the rostrum, he named it Ziphius cavirostris. From the condition of the bones, and the general characters of the specimen, he judged it to be a fossil.

When Turner compared the skull with those of other beaked whales that had recently stranded on the Shetland Islands (north of Scotland), he realized that Cuvier's "extinct" species was not a fossil at all, but similar to the Sowerby's whales of the Shetlands.

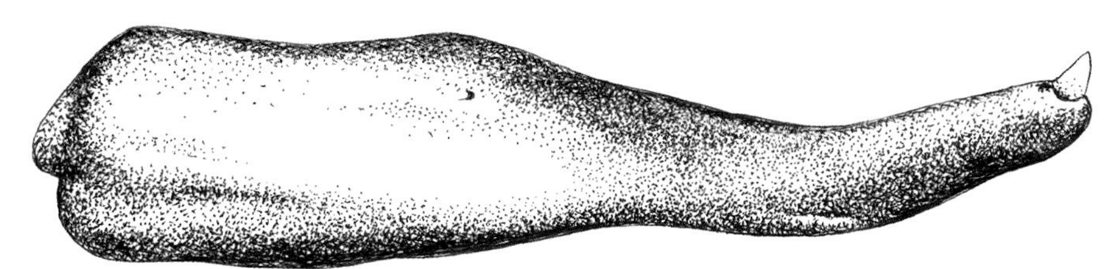

Mandible of *Ziphius cavirostris*.

The generic name *Ziphius* means "sword" in Greek (although it is usually spelled *Xiphias*, as in *Xiphias gladius*, the broadbill swordfish) and refers to the sharply pointed rostrum. *Cavirostris*, from Latin *cavus* ("hollow") and rostrum ("beak"), describes the hollow above the rostrum in the skull of adult males. This hollow deepens with age until, in old males, it extends to the palate. The vernacular name "goosebeak" refers to the shape of the beak seen in lateral view. As in the case of most other beaked whales, the males of this species have exaggerated premaxillary crests, asymmetrically enlarged on the right side of the skull. According to Tomilin (1967), "Cranial asymmetry [is] expressed very distinctly. . . . on the right side, the maxillary, premaxillary, nasal bone and occipital condyle are developed to a much greater degree than on the left. Right nasal and premaxillary bones [form] the greater part of the bony roof over the nostrils." In Japan the goosebeak whale is known as *Akabo-kujira*, or "baby-face whale," probably a reference to the small mouth and large, dark eyes.

The most frequently observed beaked whale, the goosebeak can be recognized by its sloping forehead, relatively short beak, and large size: mature individuals may be as long as 28 feet (8.4 m), and weigh upwards of 6,600 pounds (3,000 kg). They are infrequently seen at sea, but they strand more often than other beaked whales (the Smithsonian Cetacean Distributional Database counted 23 California records between 1945 and 1965), and they have often been seen breaching. We have 1,040 confirmed stranding records of this species. The color is quite variable and may be a function of age or sex, but this is still unknown. Descriptions of the animals range from dark purplish brown to slate gray and to fawn. Fraser (1946) reported that of the five specimens stranded on British coasts from 1933 to 1937, "four of them were described as being more or less white." In mature individuals, the head and back are often white, and the whales are invariably marked with a profusion of oval spots and numerous linear scars. Many observers have commented on the reversal of the normal countershading, in which the animal typically is dark above and light below, to offset the dark appearance caused by shadows on the lower surface. The goosebeak is often light on the dorsal surface and dark gray on the undersides (Harmer 1927), one of the few cetaceans—and one of the few mammals of any kind—to be colored this way. Most animals described have had dark flippers, and the underside of the tall flukes is light, sometimes even white. A living specimen was described by Aitken (1971) as "purple black on the dorsal half of the body and around the head, tail and pectoral fins, grading through dark gray-brown on the lower side to pale gray-brown on the belly."

The lower jaw extends well beyond the upper, and in males an apical pair of teeth can be seen in the lower jaw even when the animal's mouth is closed. The teeth of mature males are large and massive, but those of females, which are usually hidden below the gums, are slender and sharply pointed. Not surprisingly, this sexual dimorphism originally led to the separation of males and females into different species. True (1910) showed that the species previously known as *Hyperoodon gervaisi* was actually the female of *Z. cavirostris,* and in the same study he also showed that the previously distinct *Ziphius semijunctus* and *Z. grebnitzki* are synonymous with *Z. cavirostris.* In younger specimens, there are some thirty-three or thirty-four smaller teeth embedded in the gums, but these do not erupt and often remain undiscovered in mature adults. Tomilin (1957) wrote that the "single nonfunctional pair of teeth and the rostrum structure are a normal adaptive feature of the teuthophagi [squid-eaters]." McCann (1974) described the dietary adaptations in a somewhat more colorful manner: "The food consists largely of squids. The tongue and palate are sufficiently roughened with small papillate projections to secure the soft, slithery textures, such as cephalopods." In one report (Hubbs 1946), a stranded specimen was described as having a "broad olive-gray tongue." Most species of beaked whales are characterized by the absence of a median notch in the tail flukes, but the goosebeak often has a small one. During a 1976 whale-marking cruise, Miyazaki and Wada (1978) observed goosebeaks northeast of New Guinea, one of the few observations of living specimens up to that time. Since then, however, Robin Baird and his colleagues from Cascadia Research in Olympia, Washington, have recorded and photographed hundreds of encounters with goosebeak whales off Hawaii: "Since Cuvier's beaked whale, *Ziphius cavirostris* is too nervous about the vessel and always dive into the sea for 30–40 minutes before approaching, it is difficult to identify the species by sighting. During their jumping above the sea, however, Cuvier's beaked whales were identified by body size, the position of the dorsal fin, and the shape of the head. Cuvier's beaked whales were observed four times. Two sightings were of single animal and three whales were found in each of the other two pods."

The goosebeak is found in all oceans of the world, except the high Arctic and Antarctic. Records of strandings include Japan, the Aleutian Islands, Hawaii, Midway Island, the western coast of North America from Alaska to southern California, England, Scotland, France, Spain, Italy, and the eastern coast of the United States from Cape Cod to Florida and the Caribbean. In the Southern Hemisphere, the species has been recorded from the Cape of Good Hope, Australia, Tasmania, New Zealand, Brazil, Argentina, and Tierra del Fuego. It is usually found in waters deeper than 3,280 feet (1,000 m), and examination of captured specimens indicates that males mature at a length of 17.5 feet (5.33 m) and females at 18 feet (5.5 m). Calves are believed to be born at a length of about 7.5 feet (2.3 m), based on the examination of a 7-foot fetus that was nearly full term (Omura et al. 1955).

In 1956, a juvenile female was captured after stranding on Pebbly Beach, Santa Catalina Island, California. She was brought to Marineland of the Pacific and placed in a large tank. At first she swam slowly and seemed to be adjusting to captivity, but by the next morning "she began circling rapidly, rubbing her flukes against the tank walls. Soon she went into a frenzy, lashing the entire tank into a froth. Then she swam the length of the tank at high speed and crashed into the far wall with a thud that was felt throughout the building. A few moments later she was escorted from the tank into the connecting flume, where she died. Her charge against the wall had broken her lower jaw" (Norris and Prescott 1961). The postmortem revealed severe congestion of the lungs, which may have been a factor in the animal's stranding in the first place.

In 1947, when the eminent whale biologist Victor Scheffer examined three beaked whale carcasses in the Aleutians, he found that one was *Ziphius cavirostris* and two were Baird's beaked whales. After identifying the skulls, Scheffer and Karl Kenyon cleaned the skulls and sent them to the U.S. National Museum in Washington. Several years later, while stationed in Alaska, biologist Karl Kenyon heard that two rare whales had stranded in the Aleutians, and he headed for Amchitka Island to examine them. They were a male and female Cuvier's beaked whale, both of which had been shot. Kenyon (1961) noted that "death had been caused by a bullet which entered the left side of the head, piercing the lower jaw and throat," and that the second whale, "presumably shot at the same time, was also shot in the throat," but made no other comment on the way the whales had died. It appears that fishermen (or maybe hunters) had no compunctions about shooting strange-looking whales that surfaced within rifle range.

Even though it was regarded as one of the most widely distributed of all the beaked whales, Cuvier's was long considered a rarity for scientists and whale watchers alike—until Robin Baird started looking for them in Hawaiian waters, that is. From 2002 onward, Baird and his colleagues spent 303 days at sea, covering over 35,000 kilometers of trackline. On average, they encountered beaked whales—almost always Cuvier's and Blainville's—about once every four days. In his 2009 discussion of beaked whale research in Hawai'i, Baird wrote, "Despite our low encounter rate [lower than it might have been for bottlenose dolphins, say, but incredible for beaked whales], eight years later, more is now known about the Hawaiian populations of both Blainville's and Cuvier's beaked whales than most populations of whales and dolphins anywhere in the world." From suction-cup time-depth recorders attached to the whales, they learned that "both species feed . . . very deep in the water column, regularly diving below 1,000 meters [3280 feet] and as deep as 1,599 meters [5,214 feet] for dives that last over an hour (as long as 94 minutes for one Cuvier's) and they perform these foraging dives day and night." Between long dives, the whales make a succession of thirty to forty short, shallow dives to replenish their oxygen supply. (Observations of sperm whales, among the mammalian world's deep-diving champions, reveal the same alternating long and short dive sequences.) Calculating the dive time and depth for this species in the waters of California, Schorr, Falcone, et al. (2011) recorded a maximum depth of 2,992 meters (9,890 ft) and (for a different dive) a maximum duration of 137.5 minutes—more than two hours. This puts *Ziphius* near the top of the list of the world's best deep-diving cetaceans.

Ziphius seems to be particularly susceptible to "attacks" by cookiecutter sharks. (Is it *Ziphius*'s light color under water?) Therefore, Baird and his colleagues were able to recognize individual whales by analyzing the photographs (mostly taken by Dan McSweeney) that showed the pattern of bite marks. From these photo-identifications, the researchers learned that there was a single, widely-dispersed Hawaiian population of Cuvier's and two separate populations of Blainville's beaked whales. (Incidentally, Pearl Harbor is a major U.S. naval base on the island of Oahu and, writes Baird, "it is regularly the site of naval exercises involving mid-frequency active sonar, the same sonar that has been implicated in a number of atypical mass strandings of beaked whales elsewhere.")

ABOUT BEAKED WHALES

Classification

As of 2015, there were ninety species of whales and dolphins recognized. The largest family was the Delphinidae (oceanic dolphins), with thirty-six species. The second-largest was the Ziphiidae (beaked whales), with twenty-two species. The largest genus was *Mesoplodon*, a beaked whale genus, with fourteen species, as compared to *Balaenoptera* (rorqual whales) and *Lagenorhynchus* (a genus of oceanic dolphins), which have six species each. You can easily see that the beaked whales are the second most diverse family, while *Mesoplodon* is far and away the most diverse genus of cetaceans (Mead and Brownell 2005).

The definition of *species* has been controversial (and largely unresolved) at least since Darwin incorporated the word into the title of his 1859 treatise. Prior to Darwin, naturalists saw species as general types, which could be exemplified by an ideal specimen bearing all the traits ascribed to the species. In *On the Origin of Species*, Darwin wrote that species are precisely what they appear to be: concepts which are useful for naming groups of interacting individuals. "I look at the term species," he wrote, "as one arbitrarily given for the sake of convenience to a set of individuals closely resembling each other . . . It does not essentially differ from the word variety, which is given to less distinct and more fluctuating forms. The term variety, again, in comparison with mere individual differences, is also applied arbitrarily, and for convenience sake."

Until fairly recently, animal species were determined on the basis of physical characteristics, fossil and evolutionary substantiation, geographical distribution, and interbreeding capabilities. Nowadays, most biologists define species as "populations of organisms that have a high level of genetic similarity." This may reflect an adaptation to the same niche and the transfer of genetic material from one individual to others through a variety of possible means. In the study of sexually reproducing organisms, where genetic material is shared through the process of reproduction, the ability of two organisms to regularly interbreed and produce fertile offspring of both sexes is generally accepted as a simple indicator that the organisms share enough genes to be considered members of the same species. Thus a "species" is a group of interbreeding organisms. This definition can be extended to say that a species is a group of organisms that could potentially interbreed, even if they live in different locations, as long as they could interbreed were they ever to come into contact with each other. Unfortunately, this occasionally fails with cetaceans, as there are several interspecific hybrids like those reported between two pilot whale species (Miralles et al. 2013). Most textbooks follow Ernst Mayr's definition of a species as "groups of actually or potentially interbreeding natural populations, which are reproductively isolated from other such groups." This is known as the "biological species concept."

Molecular biology overlaps with other areas of biology and chemistry, particularly genetics and biochemistry, and is chiefly concerned with understanding the interactions between the various systems of a cell, including the interactions between DNA, RNA, and protein biosynthesis. Some biologists may view species as statistical phenomena, as opposed to the traditional idea, with a species seen as a class of organisms. In that case, a species is defined as a separately evolving lineage that forms a single gene pool (the "phylogenetic species concept"). Although properties such as DNA sequences and morphology are used to help separate closely related lineages, this definition has fuzzy boundaries. The exact definition of the term "species" is still controversial, and while biologists have proposed a range of more precise definitions, the definition used is a pragmatic choice that depends on the particularities of the species under discussion.

There is a long history of poor or incorrect identification of beaked whale species because almost all of them are inadequately represented in museum collections; many of them are similar in appearance to one another (especially juveniles and adult females); and at-sea viewings of living animals are few and far between. In order to understand ziphiid speciation, Meryl Dalebout and four colleagues developed "A Comprehensive and Validated Molecular Taxonomy of Beaked Whales," published in the *Journal of Heredity* in 2004. "Samples were obtained from stranded or beach-cast animals, victims of incidental fisheries (by-catch) and museum collections . . . and reference sequences were generated only from validated specimens. By making the beaked whale mtDNA database accessible to researchers everywhere, at least part of the problem of identifying the various species—old or new—will be solved."

There is no accepted standard for listing species in a particular family or genus. The family Ziphiidae, erected by J. E. Gray in 1850, includes the genera *Ziphius* (one species)*; Berardius* (two species); *Tasmacetus (*one species), *Indopacetus* (one species), *Hyperoodon* (two species), and *Mesoplodon* (fifteen species). In his 1998 comprehensive (and authoritative) discussion of the systematics and distribution of the marine mammals of the world, Dale Rice wrote, "This list . . . like any such list, is only a progress report—a synopsis of our knowledge and uncertainties at the time it was written. No list of scientific names can ever be considered the 'correct' list. Taxonomists sometimes disagree with each other, and classifications are changed continually as new facts are brought to light and new interpretations emerge."

Mesoplodon—which translates as "armed with a middle tooth"—is the generic name of a group of whales that can be found (but not often) in all the temperate and tropical oceans of our planet. The skull—often the only part examined—has such elongated nasals and mandibles that it resembles the skull of a gigantic bird. Baird's beaked whale has been measured at 41 feet (USNM 49725), but average length for the 22 recognized species is about 20 feet.

The name *Indopacetus pacificus* (which can be translated as "whale of the Indo-Pacific") is probably a good place to introduce and explain the concept of scientific nomenclature. A century before Darwin published *On the Origin of Species*, the Swedish naturalist Carolus Linnaeus (1707–78) developed a system of classifying every living thing, giving names to more than four thousand species of animals and nearly eight thousand kinds of plants. To do this, he invented the system of binomial nomenclature, still in use today, where the first of two names (the genus) identifies a group of related species, and the second, known either as the specific or the trivial name, identifies the species within that genus. You belong to the genus *Homo,* and you are of the species *sapiens,* making your binomial *Homo sapiens.* But because you belong to the only living species in the genus *Homo,* this does not make for a particularly instructive example of taxonomic diversity. Here's a better one: the genus *Mesoplodon* includes most of the beaked whales, but some differ enough from the mesoplodonts to have been assigned a different generic name: *Berardius, Hyperoodon, Ziphius, Tasmacetus,* or *Indopacetus.*

Common names vary from language to language; beaked whales are called *ballenas pico de botella* (bottle-beak whale) in Spanish, *baleines à bec* (whales with beak) in French, and so on, and therefore the scientific names are of critical importance for an unambiguous understanding of which animal is being discussed. To insure that everyone is on the same page, you need

Mesoplodon densirostris or *Hyperoodon planifrons*, not just "dense-beaked whale" or "Southern bottlenose whale." (Scientific names always appear in italics, and the generic name always begins with a capital letter, the specific name with a lowercase initial letter. For convenience in binomial nomenclature, when listing members of the same genus, the generic name can be represented by its initial, as in "*M. europaeus*," "*M. densirostris*," etc.).

A suite of anatomical characteristics differentiates the beaked whales from all other cetaceans: reduced dentition, elongate rostrum, paired throat grooves, small flippers, fluke notch reduced or absent, and skull characters. Any cetacean with all these characters is a beaked whale, but the twenty-two kinds are all different from one another, otherwise they would not be separate species. For each one, the arrangement of characteristics defines the species. From the teeth, you can identify the male *Mesoplodon layardii, M. densirostris, M. ginkgodens,* and other species, but what about the females whose teeth are not visible? The proportions and coloration are different for each species, but the color of a stranded whale often changes dramatically if it has been baking in the sun for a while. In his *Whales and Dolphins of New Zealand and Australia: An Identification Guide,* Alan Baker emphasizes those elements that are not likely to change unless the animal has completely decomposed: silhouette, forehead bulge, beak length, teeth (if applicable), and again if applicable, coloration.

And now, let's say something about echolocation signals. In a 2013 discussion in the *Journal of the Acoustical Society of America,* Simone Baumann-Pickering and ten colleagues combined data collected in various locations and were able to identify the frequency-modulated upsweep pulses recorded from various beaked whale species. For the most part, the echolocating whales had first been identified by surface sightings by researchers whose observations had been previously published. The signals recorded from Baird's beaked whale (Southern California); Longman's beaked whale (Hawaii); Blainville's (Hawaii and Saipan); Gervais' (Gulf of Mexico); Stejneger's (Aleutians, Washington, Southern California); and Cuvier's (Aleutians, Hawaii, Southern California) were *species specific,* meaning that the species could be identified by the differences in their signals. It is still a subject for research as to why these echolocation signals are different. Are they matched to the beaked whales' prey or some other aspect of their diving behaviors?

The 2008 *Marine Mammals of the World* by Jefferson et al. is subtitled "*A Comprehensive Guide to Their Identification,*" and the book might conceivably be brought aboard a whale-watching cruise. It would help in differentiating the subspecies of Pacific spinner dolphins and identifying Commerson's dolphin if your ship happened to be around Tierra del Fuego, but although there are keys for identifying the various beaked whales, you're simply not likely to see one. And if by some miracle you did, you wouldn't be able to see the teeth clearly enough to know one species from another. The teeth, not used for eating like most teeth, are particularly useful to scientists in the identification of species—in males, anyway—as they differ from one species to another in size, location, and function. The so-called saber-toothed whales have a pair of large teeth rising from the lower jaw, probably used against rival males in nonlethal battles; other species have smaller teeth at the point of the lower jaw, the function of which is not known (unless they're simply an evolutionary holdover from the time the ancestors of beaked whales had proper teeth); one species has a mouthful of teeth like a dolphin, and one has teeth so strange that the owner can barely open his mouth.

Beaked whales can tuck their flippers into depressions on their sides that are known as "flipper pockets." The tucked-in flippers probably reduce drag, particularly when the animals are diving. (The bluefin tuna, among the most streamlined of all fishes, can tuck its pectoral fins into indentations along its flanks and its first dorsal fin into a slot on its back.) Mead, Walker, and Houck (1982) list *Mesoplodon carlhubbsi*, *M. ginkgodens*, *M. stejnegeri*, *M. densirostris*, *M. europaeus*, *M. mirus*, and *Ziphius cavirostris* as those beaked whales with pockets, but in their guide to the marine mammals of the world, Jefferson, Webber, and Pitman (2008) say that all species have these indentations. A less pronounced arrangement can be seen in sperm whales, but not in cetacean species that spend most of their lives at or near the surface. The pockets must have something to do with deep diving.

Aside from weird teeth, body scarring, flipper pockets, suction feeding, bizarre coloration, stranding propensities, and anomalous distribution patterns, one of the more remarkable things about the beaked whales is their taxonomy, wherein new species are added, old ones dropped, and the names changed to reflect centuries-old descriptions or surprising new discoveries. They are among the few large animal groups where the literature is in such a continuous state of flux, correcting misidentifications as new (or old) material becomes available or deciding that "new" species are really older ones after all. Here's an example: In a 1999 discussion of new records of beaked whales from New Zealand, Alan Baker and Anton van Helden wrote that "a reference in Baker (1983) to a record of *M. ginkgodens* from the Chatham Islands to the east of New Zealand, was based on a tooth whose size and proportions are almost identical to the type (shaped like the leaf of a ginkgo tree).

Subsequent study of a series of mesoplodont teeth in New Zealand and in the United States has shown, however, that the Chatham Islands tooth is a large male *M. grayi*, whose teeth differ from those of *M. ginkgodens* only [in] being slightly more asymmetrical in lateral profile." In other words, the teeth that were used to extend the range of the ginkgo-toothed whale to New Zealand actually came from the commonest species in New Zealand waters. So does *M. ginkgodens* occur in New Zealand? Yes, it does, but not with the teeth of *M. grayi*.

The taxonomic modifications, which happen fairly often, are functions of the "rarity" of these deepwater pelagic creatures that live most of their lives well out of sight of inquisitive human observers. But in fact, the beaked whales may not be all that rare—we just haven't been able to spend enough time in the middle of the ocean waiting for them to appear.

Because so many beaked whales have been identified from beached carcasses spoiling in the sun (or even skeletal remains) the coloration of the living animal has not always been available for individual species identification. In recent years, however, the outward appearance of the living beaked whales spotted at sea has become more useful as a species determinant. *Mesoplodon layardii* of the strap teeth is adorned with a black, gray, and white pattern that rivals that of the killer whale in complexity. The distinctive white head and fawn coloration of *Ziphius* sets this species apart from any other whale—and from any other mammal, for that matter.

Distribution

Another element of continuing ambiguity in the study of beaked whales is their distribution. In the introduction to a 2006 discussion of the "Known and Inferred Distributions of Beaked Whale Species," Colin MacLeod and colleagues wrote, "Information regarding beaked whale species is so sparse that even the most basic aspects of their biology, such as their distribution, remain poorly defined for some species . . . While for some species, such as the relatively commonly recorded Cuvier's beaked whale, the inferred distribution is likely to be an accurate reflection of the species' actual distribution, for lesser known species, such as the spade-toothed whale, the inferred distribution is more tentative."

On a world map for each species, the authors plotted specific locations, "based on a review of published information and from unpublished sighting and stranding records by the authors or obtained from other sources" but then admitted that "a lack of specific point locations should not be interpreted as a lack of occurrence of a species in a specific area," because dead or incapacitated animals might be transported by ocean currents to locations far from their usual range and also because "species ranges are not static, and can change over time." Trying to pinpoint beaked whale occurrences in the world's oceans leaves enormous swaths of ocean unmarked, which does not mean that beaked whales do not live there—it only means that they have not been recorded there.

Beaked whales are so rarely spotted—dead or alive—that almost every sighting is noteworthy. Scientists do not often call the newsroom or local television station with the results of their research projects, unless there is something particularly unusual about their observations or conclusions, such as a rare species sighted at sea or a mass stranding that might be attributed to offshore military maneuvers. For the most part, though, the very appearance of a beaked whale, on the beach or in the water, brings out the cameras, and the local reporters file a story about a "rare kind of whale found on the beach." If you subscribe to "Google Alerts" on a particular topic, your e-mail inbox will receive regular notifications related to the topic you have selected. From January through October 2013, we received alerts consisting of news stories from around the world where the words "Beaked Whale" were in the title. The alerts during that ten-month stretch included a ginkgo-toothed whale and a Blainville's beaked whale (both dead) in the Philippines; Shepherd's beaked whales (live with photo) in Australia; an Arnoux's beaked whale (dead) in Cape Town; a Cuvier's beaked whale (dying) in New Zealand; a True's beaked whale (dead) in New Zealand; a spade-toothed whale (dead) in New Zealand; a Sowerby's beaked whale (dead) in Wales; a Cuvier's (dead) in California; a Sowerby's (dead) in Ireland; a Cuvier's (filmed at sea) off Oman; a Cuvier's (dead) from New Zealand; a Shepherd's (filmed at sea) off South Australia; a Baird's (killed by Japanese whalers); a Gray's (dead) from New Zealand; a Gervais' (dead) in Florida; a True's (dead) in South Africa; a Sowerby's (dead) from Scotland; and Baird's (mother and nursing calf, dead) in Valdez, Alaska. Many other beaked whales probably stranded on remote beaches where there was nobody to find or report them, but these "alerts" demonstrate how rare and newsworthy they really are. Any and every report of a beaked whale adds something, however small, to our knowledge base of these enigmatic but endlessly fascinating animals.

Why do we need to know the distribution of the beaked whales? For a better understanding of the lives of these remarkable creatures, of course, but also because we have begun to realize that not all strandings have been "naturally" motivated. We still have very little information on what causes whales to beach themselves, but there is no doubt that high-powered naval sonar has caused the stranding deaths of beaked whales at locations around the world. In order to prevent these disasters from recurring, we have to have a better idea of where the whales live.

The State of Our Knowledge

The beaked whales are still poorly known. The introduction to a 2002 article (Dalebout, Mead, et al.) introducing *Mesoplodon perrini*, a new species of beaked whale that brought the actual total to 22, read as follows:

> Beaked whales are the least known of all cetacean families with 21 species currently recognized, they are second only to oceanic dolphins with 35. Beaked whales are rarely observed at sea due to their preference for deep ocean waters and elusive habits. Most species are known from only a small number of stranded specimens, and several have never been seen alive. Of the twelve cetacean species described in the last 100 years, eight have been ziphiids, primarily of the genus *Mesoplodon*. In the closing decade of the 20th century, two new beaked whales were discovered; the lesser beaked whale *M. peruvianus* (Reyes et al. 1991), and Bahamonde's beaked whale *M. bahamondi* (Reyes et al. 1995) although the latter is now recognized as synonymous with *M. traversii*.

Robert Pitman, a National Marine Fisheries Service marine biologist, has spent the past thirty years on a cetological quest: to see every living cetacean species in the wild. In a letter dated December 2012, as he was leaving for a month in the Antarctic, he wrote,

> I have for some time (decades!) been on a mission to see all of the world's cetaceans. I just got back from 3 weeks in West Africa where I went to see Atlantic humpback dolphin, which was my 80th cetacean. Of the 87 currently recognized species I have yet to see Bolivian River dolphin (a recent split; easy, but it means I have to go to land-locked Bolivia to see my last dolphin species); Omura's whale (New Caledonia seems to be the best place but I don't think this species has been identified alive in the wild—I'll need to photograph and biopsy the animal to be sure). The final 5 species are of course *Mesoplodon*—and all species that to my knowledge have never been identified alive in the wild. I have done fairly well on the beaked whales though—both *Berardius*, both *Hyperoodon*,

> *Ziphius*, *Indopacetus*, *Tasmacetus* and 9 species of *Mesoplodon*. Some people collect stamps . . .

That Bob Pitman, a man who has devoted his entire career to accumulating a life list of cetaceans, is still missing 5 of the 22 beaked whales, tells you everything there is to know about *Mesoplodon*.

In his 1928 discussion of the evolution of whales (when there were thought to be only ten mesoplodonts), Remington Kellogg stated,

> inasmuch as there are ten recognized living species of the genus *Mesoplodon*, there can be no objection to the assumption that species of these ziphioids were equally if not more numerous in preceding geological periods . . . Unfortunately the majority of fossil specimens referred to this genus consist solely of rostra, which are subject to age, individual, and sexual variations. Some of these specimens differ very little from living species, and one of them, *Mesoplodon longirostris* (Cuvier 1836) may have been a precursor to the living beaked whales, *Mesoplodon bidens* and *M. europaeus*.

The fossil record may be sparse because of the deep-water habitat of these whales—if they happened to die at sea, the carcasses might sink to the bottom in thousands of feet of water. But as we shall see, offshore bottom-fishing trawlers have begun bringing up remnants of long-extinct—and completely unanticipated—species of beaked whales that are redefining what we thought we knew about these enigmatic creatures.

Teeth

Except for that singular tooth in narwhals, there is no stranger dental dimorphism than the teeth of beaked whales. Males have erupted teeth, females of most species do not. Most species have their only two teeth in the lower jaw. Some have a pair of teeth (one on each side of the jaw). Some have four teeth in the lower jaw, two big ones in front and two smaller ones just behind them. Some have inexplicable grooves on the leading edges. Beaked whale teeth are among the strangest in the animal kingdom. *Mesoplodon layardii*, the strap-toothed whale, has two teeth that grow from the lower jaw, up and over the upper jaw to form an arch that nearly prevents the animal from opening its mouth. The "reason" for this apparently counterproductive development is a mystery.

Other *Mesoplodon* species have teeth that are not quite so self-defeating but just as unusual. One species, discovered in Japanese waters in 1958, has two teeth that scientists likened to the leaves of the ginkgo tree, so they named the species *Mesoplodon ginkgodens*. In most species of *Mesoplodon*, the two teeth of the male are located at about the midpoint of the lower jaw—hence the name—but some of these teeth are beyond extraordinary. The Bering Sea beaked whale, *Mesoplodon stejnegeri* (named for its discoverer, Leonhard Stejneger), has teeth that can be four inches wide and only half an inch thick. *Tasmacetus* is the only species of beaked whale with a whole set of its teeth inside the mouth, where you might think they belong, but this species (males and females) also has two big teeth at the apex of the lower jaw in males that protrude even when the mouth is closed. These teeth do not erupt in females. Those beaked whales with two big teeth in the middle of the lower jaw that are visible outside the upper jaw when the mouth is closed are known as saber-toothed whales. From the examination of the stomach contents of stranded specimens, it is known that the staple of the beaked whale diet is squid, but the teeth, whether two or four, cannot be of much use in obtaining food, since females and juveniles, with no teeth at all, manage to feed themselves quite satisfactorily.

The teeth of adult males are diagnostic in identifying the species, but females present more of a problem. Of course the range, coloration, and size are helpful in species identification, but where similarly sized females share a habitat, it is very difficult to tell them apart. (Nowadays, DNA analysis is a great help, but prior to its introduction, identification of juveniles and females was a problem.) There is a case where a toothless female was classified as a distinct species and christened *Aodon* ("without teeth"). In Robert Hamilton's 1839 *Naturalist's Library (Mammalia, Whales, Etc.)*, we read: "This elegant whale belongs to the genus *Aodon*, which is characterized by having no baleen and no teeth . . . The specimen now before us was stranded in 1825 near Havre; was examined by Blainville, Dr. Suriray and the son of M.F. Cuvier, and the skeleton deposited in the Paris Museum." In his 1900 *Book of Whales* Frank Beddard wrote, "It was kept out of the water for two days before it died. It was fed on 'soaked bread and other alimentary substances,' and it was heard to emit "a low cavernous sound like the lowing of a cow." Beddard further commented, "The toothless whale of Havre, it may be remarked, named *Aodon dalei*, seems to be merely a toothless, probably aged, example of *Mesoplodon bidens*."

Although the number, location, size, and shape of the teeth of adult male beaked whales was long a mystery (they were certainly useful to biologists for differentiating species) it was not immediately evident how the whale might put such a strange dental arrangement to use. In his 1984 study of "fighting" between adult male *Mesoplodon carlhubbsi*, John Heyning suggested that the teeth of this species (and by extension, most other beaked whale species) are used for ritual fighting, which inflicts nonlethal wounds on opponents. No human has ever witnessed this activity; it is only deductively that we can explain the interconnected dentition and the extensive body scarring on adult male beaked whales. Given the depth and inaccessibility at which these encounters take place, we have only circumstantial evidence to show that they occur, but the evidence is fairly conclusive. In most species, only the males have erupted teeth, usually oddly shaped, tusk-like protrusions from the lower jaw.

Scarring and Fighting

We know that suction feeding does not require teeth, so we can safely assume that the teeth of beaked whales are not used for capturing slippery food items, as are the teeth of dolphins. As the males of ziphiids are often extensively scarred, we might conclude that they do battle with one another with their mouths closed, bringing into play the protruding tusks, which leave parallel tracks on the head, back, and sides of the opponent. (Females of the various species have no protruding teeth—and no scars.) As Heyning (1984: 1645) put it:

> Aggressive social interactions involving the use of teeth are noted in several groups of odontocetes [toothed whales]. This includes both direct observations in the field and in captive situations, and indirect morphological clues such as external scarring patterns and tooth damage . . . This is due to the delicate nature of cetacean epidermis which is easily injured and often scars a contrasting color to the surrounding skin. Owing to their apparent rarity, there are no detailed accounts of beaked whales fighting. However, the extensive scarring patterns seen on their bodies indicate that intraspecific fighting does occur.

In 1998, Colin MacLeod, the author of the "species recognition" hypothesis of jaw shape, published an essay wherein he attempted to document his theory that body scarring was an indicator of male "quality" in aggressive social interactions: the more heavily scarred an individual, the more powerful he is, and therefore the more attractive to females. MacLeod compared the

Two northern bottlenose whales photographed in 1990 by the Whitehead Lab just below the surface, showing that they sometimes swim upside down, making the dorsum-to-dorsum scarring behavior at least possible.

body scarring of beaked whales to that seen on Risso's dolphins (*Grampus griseus*), 10-foot-long dolphins that are always heavily scarred and also have teeth only in the lower jaw. The extensive scarring on male and female Risso's dolphins has never been satisfactorily explained.

In this 1998 paper, MacLeod does not mention the possibility, which he raises later, that extensive scarring on the backs of male large-toothed beaked whales would have to be inflicted with one of the whales swimming belly-up to enable the teeth to be brought into play. But in his 2000 discussion of the teeth in Sowerby's beaked whale, MacLeod describes parallel, longitudinal scars on the dorsal surface, "consistent with the hypothesis that males approach and pass each other dorsum to dorsum"; and in his 2002 discussion of *M. densirostris*, he repeats the hypothesis, accompanying it with a photograph of a heavily-scarred individual and a drawing of the male-male conflict to support his arguments.

Could it be that beaked whales rely on echolocation to calculate the position of their opponent? When we asked Olivier Lambert, a beaked whale paleontologist, about this possibility, he said, "If beaked whales are able to locate and catch squid with a mantle length no more than 30 cm in total darkness at 800–1,000 meters of depth, they can surely detect an opponent a few meters long. Maybe the fights are more ritualized in some species, with only minor contacts, but still leaving scars." Of the twenty-two species of beaked whales, only a few have protruding tusks that might be used to inflict parallel scars on their opponents. *Mesoplodon densirostris, M. carlhubbsi, M. stejnegeri, M. traversii, M. bowdoini,* and *M. peruvianus* might be able to perform these stunts.

In a letter written in September 2013, Alan Baker dismissed MacLeod's "dorsum-to-dorsum" explanation of the scarring, because many of the scars are not parallel (as they would have to be if one whale inflicted the scars with both teeth simultaneously). They often show switchbacks or turns that are commensurate with the hypothesis that one whale approaches the other from the side and strikes his opponent a glancing blow with only one of his protruding teeth. (This deployment also explains how whales without grossly protruding daggers, such as species like *M. grayi* with smallish teeth located at the midpoint of the jaw, can inflict single, random scars on their rivals.) Male battles where one of the adversaries charges his opponent from the side and rakes him with one tooth before swerving off makes much more sense than a belly-up whale approaching his opponent blindly and hoping that his opponent will allow himself to be hit (just as the attacker can sense the position of the victim, the victim can sense what's coming). The "face-to-face, glancing blow" hypothesis puts beaked whales comfortably into the camp of other male-male dominance conflicts, comparable to sheep, deer, goats, and antelope (and even elephant seals), without making the already strange beaked whales appear even stranger.

There is little doubt that the linear scars on the bodies of beaked whales are somehow caused by other beaked whales, but the little round scars are caused by a two-foot-long sharks known as *Isistius brasiliensis*. These little sharks attack swordfishes, marlins, dolphins, pinnipeds, and even the larger whales, by taking little circular bites out of their flanks—which accounts for their common name, "cookiecutter sharks." The cookiecutter has small erect teeth in the upper jaw and large triangular teeth in the lower, so it affixes itself to its prey and then spins on its axis to remove a cookie-shaped plug of flesh.

Mackintosh and Wheeler (1929: 373) first discovered these scars on blue and fin whales taken at South Georgia Island. The source of these "crater wounds" was long a mystery, but in 1971 biologist E. C. Jones solved the mystery by holding the open mouth of a dead *Isistius* up to a nectarine and twisting it as he imagined the shark would do as it bit a larger animal. The result was perfect plug-like core removed from the fruit, exactly like the piece that would have been removed from the living victim. (These sharks evidently prey—or attempt to—on nonliving victims as well, as evidenced by the appearance of the same bites on the rubber coating of the sonar domes of submarines.) Cookiecutters do not differentiate between male and female beaked whales, so little round scars are found on almost every specimen. When completely healed, they show up as light-colored spots, about the size of a lemon. The random pattern of the scars can be used to identify individuals. In his 2009 discussion of beaked whale research in Hawaii, Robin Baird wrote: "Because of the large number of scars caused by cookie-cutter sharks, we are able to recognize almost every individual we encounter using photographs."

In the early days of beaked whale research, all white oval scars were reported as lamprey scars. Gordon Pike, in 1951, described actual lamprey scars. He recognized the difference between his British Columbia scars, some of which bear tooth marks of lampreys, and the scars described by Mackintosh and Wheeler (1929), which turned out to be cookiecutter. We recognize that most of the scars are caused by cookiecutter sharks, but some scars on beaked whales turn out to be actual lamprey scars.

The dense ossification of the rostrum of male beaked whales may also be useful in fights between males. It was first recorded by French zoologist Henri de Blainville (1777–1850) in his 1817 description of the beaked whale whose specific name—*densirostris*—is a literal reference to the dense bone of the beak. (Another name for this species is Blainville's beaked whale.) Other species, such as *Ziphius cavirostris* and *Mesoplodon carlhubbsi*, share this hyperossification, but not everybody agrees that the dense rostrum is used for ramming rivals; it could possibly serve as an aid in echolocation.

Feeding

After carefully dissecting and analyzing the heads of various mesoplodonts, Heyning and Mead (1996) found several unusual structures that supported the theory that beaked whales were obligate suction feeders: The tongue could be retracted, and the hyoid bones were greatly enlarged to support the muscles that control the tongue, enabling the animals to create low pressure in the mouth, which permits water to rush in—the very definition of suction. An examination of the muscles around the mouth revealed that "Ziphiids are unable to open their mouths very wide . . . the combination of the small gape and tissue around the corner of the mouth results in a relatively small and somewhat restricted oral opening." Finally, when the stomach contents included squid without puncture marks, it became obvious that beaked whales did not use their teeth to capture their prey. The throat grooves on mesoplodonts were shown to be distensible, contributing to the suction-feeding hypothesis.

The depth and remoteness of their habitat means that (so far) no one has ever seen an adult beaked whale feeding. We do know that harbor porpoises can suck in a fish "with some force" from the hand of a trainer; and walruses, which are not related to whales, beaked or otherwise, feed almost exclusively by suction, slurping clams out of their shells on the sea bottom. But the most significant element in the suction theory is the observation of live-stranded beaked whale calves that were temporarily held in captivity and in which suction feeding was believed to be observed.

In discussing the northern bottlenose whale (*Hyperoodon ampullatus*), Alan Baker (1983) wrote that it "is known to dive to great depths in search of food, and it is thought that the massive fatty forehead acts as an acoustic lens for directional beaming of echolocation signals, and prey stunning sound bursts." This is one of the few suggestions that at least some beaked whale species might beam sound bursts at their prey to immobilize or even kill them before sucking them up. Norris and Møhl (1983) speculated that sperm whales disable their prey in this way, but at the depths where this might occur, no human being has ever witnessed it—or even knowingly heard it. Without these debilitating sound bursts, however, there is a gap in the feeding sequence for sperm whales and beaked whales. Once the whale has located the prey, and before it can slurp it up, it has to be able to slow it down or even kill it. Because the whales are air breathers and are holding their breath, while the squid are water breathers, the whales cannot be dashing around in the dark hoping to bump into a swimming squid. Beaked whales cannot open their mouths very wide, so there has to be a way for a feeding whale to get enough squid into its gullet to make an hour's deep dive worthwhile. (In their 1996 analysis of the diet of strap-toothed whales, Sekiguchi, Klages, and Best found that males and females manage to eat the same size squid.)

In deepwater locations as far apart as the eastern Mediterranean and submarine canyons off eastern North America, scientists in submersibles have reported mysterious "gouges" in the mud of the seafloor. Woodside et al. (2006) observed these gouges at depths ranging from 1,700 to 2,100 meters (5,575 to 6,888 ft) on the mud volcanoes around Greece and Turkey. They described them as "a central groove (about 10 cm deep and 1–2 m long [3 by 30–50 inches] superimposed upon a broader bowl-shaped depression with raised rims on either side of the central groove." Auster and Watling (2010) described the gouges in the North Atlantic canyons as "approximately 0.5–1 m long, and 20–40 cm wide," and suggested that "whale beaks may have contacted the seafloor in the process of capturing prey . . . The shape and size of the gouges causes us to suggest causation by foraging beaked whales, however, we cannot completely eliminate other different or currently unknown causes."

The beaked whale species sometimes spotted in the vicinity of the steep-sided canyons of the Atlantic are northern bottlenose whales (*Hyperoodon ampullatus*), Sowerby's beaked whales (*Mesoplodon bidens*), Blainville's (*M. densirostris*), Gervais' (*M. europaeus*), and Cuvier's (*Ziphius cavirostris*). These whales, all accomplished deep divers, have only been seen at the surface (and then rarely), so assigning these gouges to feeding ziphiids at depth is pure speculation. But as the authors offer no other plausible explanations for the "causation" of the gouges, we're stuck with an inexplicable version of beaked whale suction feeding. As Woodside and colleagues (2006) write, "The suction feeding mechanism of beaked whales would allow the selection of prey within the mud and then expulsion of mud as water held in the forestomach is released past prey secured against the rugose palate."

Many male beaked whales have a "hollow" worn into the leading edge of their tooth pair. In a 2012 letter to Richard Ellis, New Zealand researcher Anton van Helden said:

> The tooth wear that you are talking about occurs in all adult male Mesoplodonts with exposed tusk teeth, and a slightly different but equally noticeable wear occurs in Ziphius. It certainly is very apparent in *M. traversii, M. layardii, M. bowdoini, M. stejnegeri,* and *M. grayi*. I strongly believe that it is as a result of the suction feeding of squid. I suspect that it may be ammonia from the squid plus the scraping of suckers that effect this. . . . I suspect more than anything that there may be some chemical reaction involved along with the continuous movement across the leading edge. The wear is essentially only on the leading edge so that makes some sense.

It is difficult to envision a suction-feeding technique that puts captured squid in a position to wear a groove into the teeth. These enigmatic "hollows" only serve to point up how little we know about the daily lives of these cetaceans of the remote pelagic reaches.

Nocturnal land mammals often have larger eyes than their diurnal counterparts, and in some primates (e.g., tarsiers, bushbabies, galagos, and lorises) the eyes are enormous. Owls have relatively large eyes, but they also depend on hearing for finding prey. Many benthic marine predators, such as swordfish, large squid, and deepwater sharks, have huge eyes, the better to see their prey in reduced light conditions. There is a big-eye squirrelfish (*Priacanthus*) and a big-eye thresher shark (*Alopias superciliosus*), both denizens of deep (or dark) waters. The giant squid (*Architeuthis*) is reputed to have the largest eyes in the animal kingdom (as many as eight inches across), and it has been postulated that the squid needs those huge eyes to spot and avoid its traditional nemesis, the sperm whale. The eyes of a 60-foot-long sperm whale are not particularly large, because it really depends on echolocation to find its prey. So it is with the beaked whales: the small size of the eyes and the low light conditions in which they hunt tell us that they use something other than vision for hunting. That something is sound.

In the Canary Islands and the Bahamas, researchers affixed acoustic recording tags to beaked whales with suction cups and listened to the sounds produced as the whales foraged in waters that were 3,000 feet deep (Johnson et al. 2004, 2008; Madsen et al. 2007). The researchers used acoustic recording tags to detect echolocation clicks produced by Cuvier's and Blainville's beaked whales (respectively, *Ziphius cavirostris* and *Mesoplodon densirostris*) during dives of up to 1,888 meters (6,192 ft). Before the whales reached a depth of 600 feet, they were silent, but beyond that depth, they clicked almost continuously. They might not click in shallow waters to avoid being detected by shallow-water predators such as killer whales, but the deep-water clicking behavior might also be a function of the deepwater habitat of the prey species. The researchers reported that the whales "produced two distinct click types associated with different kinds of echolocation-mediated foraging. Long duration FM [frequency modulated] clicks are used during the search phase and the initial approach to prey. When the target is about one body length away, the whale switches to buzz clicks. These short duration wide-bandwith pulses are produced at a high rate throughout the capture attempt."

This is more or less what most dolphins do, but the beaked whales echolocate in an environment that is much deeper and darker. The researchers noted that the beaked whales followed each deep foraging dive with a series of shallower dives, probably to replace the depleted oxygen consumed during the deep dives. In a 2006 study titled "Extreme Diving of Beaked Whales," Tyack, Johnson, and colleagues acoustically tagged two species of beaked whales, *Ziphius cavirostris* in the Ligurian Sea (north of Corsica in the Mediterranean) and *Mesoplodon densirostris* off the Canary Islands in the Atlantic, and wrote, "Beaked whales are considered to be deep divers based primarily upon diet inferred from stomach contents [mostly squid]. This family includes some of the world's most cryptic and difficult to study mammals and little is known about their diving behavior" (Johnson, Hickmott, et al. 2008). The researchers also recorded echoes from the prey, the first time this had been accomplished for an animal echolocating in the wild. The actual movements of the whales could also be tracked, and an analysis showed that they often circled the prey while buzzing. Beaked whales can adjust their echolocation behavior and movement for capture of different prey on the basis of structural echo information. Once the prey is in range, they suck it right up.

All beaked whales are deepwater species, taking to the shallows only in times of stress. This would account for the infrequency of sightings; most species are known from a very limited number of observations, of either stranded specimens or, very infrequently, live animals at sea. As Jay Barlow and his colleagues wrote in a 2006 discussion of the abundance of the Ziphiidae, "Small beaked whales are more difficult to visually detect in the field than most other cetaceans . . . They typically surface inconspicuously, usually without a splash or visible blow, and seldom breach or display other aerial activities. In addition, small beaked whales rarely display their flukes when they dive and they occur in small groups, typically 1–5 individuals. Finally, they spend very little time at the surface, and then dive for extraordinarily long periods."

Our inability to describe the feeding habits of beaked whales demonstrates how effectively the deep ocean has helped these animals keep their secrets. Much of what we know about their lives has been learned from watching them at the surface and then speculating (or guessing) about what goes on in the depths and out of sight. The examination of carcasses and skeletons has also provided much useful information about their *modus vivendi*, but the window is still shut on how they actually use some of their exotic devices and anatomical modifications. Technology has enabled us to listen to their various sounds and track their dives with radio tags, but no one in a submersible has seen a beaked whale feed, and no camera has recorded an image of one of these unfathomable deepwater denizens capturing their prey.

From Land to Water to Tooth Loss

Millions of years ago, there existed some terrestrial mammals that would perform one of the most astonishing reversals in evolutionary history: they would return to the sea to become marine mammals. Paleontologists have unearthed a remarkable succession of terrestrial and semiterrestrial mammals that presaged today's whales. There are *Pakicetus*, a long-nosed, long-tailed mammal that looked like a giant shrew; *Ambulocetus* the "walking whale," whose appearance has been likened to a furry crocodile, and *Basilosaurus*, the erroneously named early whale (*Basilosaurus* means "king of the lizards"). *Basilosaurus* and kin form the suborder Archaeoceti from which the living cetaceans are descended.

Odobenocetops, the "walrus whale," looks remarkably like an animal predestined for extinction. Drawing by Richard Ellis after restoration by Mary Panish.

The complex history of fossil whales, far from complete or definitive, provides some signposts along the road from terrestrial to marine mammals. The "fossil record" opens a small window on what lived before; but there are surely many extinct creatures whose remains we may never find. In the oceans of the Miocene, 23 to 5 million years ago, there swam a diverse assortment of ziphiid cetaceans, similar in many ways to the beaked whales of today.

Somewhere along the line, a divergent branch of the toothed-whale lineage began to develop those characteristics that would set them apart from all other odontocetes: strange teeth positioned in strange places and used primarily for fighting, not feeding; paired throat grooves to allow for gular expansion and suction feeding; and diving capabilities unparalleled in the mammalian world. Chopped from remote rock formations in the earth's almost impenetrable geological history, and from the dark, muddy bottoms of the sea, the ancient beaked whales begin to appear to science.

We can never know with certainty the "why" of evolutionary adaptations, so biologists have to create hypothetical scenarios that can be used to explain what we see. How did so many of the beaked whales come to lose all but two of their teeth? How did species evolve paired tusks that are employed in ritual fighting and not in eating? In a 2006 article entitled "Adaptations and Evolution of Structures for Intraspecific Combat in the Family Ziphiidae," Colin MacLeod et al. wrote,

> Beaked whales can be separated into three conditions based on the form of sexually dimorphic structures and mode of intraspecific combat. The first condition is "jousting" using large, apical tusks. This is the ancestral condition in beaked whales and evolved from the "bite and rake"

> combat found in many toothed whales. "Jousting" using apical tusks risks damage to the melon, which is required for echolocation. This led to the evolution of a "shield" made of dense connective tissue to protect the melon, giving such species a charateristic rounded forehead. From this basic condition found in most genera, two derived conditions have evolved. In some *Mesoplodon* species, the tusks have been shifted into a more posterior position, reducing the likelihood of damage to the melon. As a result, in these species the melon "shield" has been reduced resulting in a more sloping forehead. In the bottlenose whales, the melon "shield" has increased in size to the point where the forehead is extremely bulbous in mature males and the "shield" itself can be used as a weapon. As a result, the mode of combat has shifted from 'jousting' to head-butting. In the northern bottlenose whale, the 'shield' is further reinforced by bony maxillary crests to make the forehead a more effective weapon.

We will never know if the early beaked whales began to lose their teeth and had to resort to suction feeding or, having resorted to suction feeding, their teeth became redundant and were reduced in number and assigned an entirely different function. Whatever the order of occurrence, some primitive toothed whales showed many of the characteristics of our living species, including densification of the rostrum. By the middle Miocene, fossil ziphiids were abundant and widespread. This was a period of maximum diversity of the Cetacea, especially for the ziphiids. In 1991, Christian de Muizon classified modern and fossil Ziphiidae into three subfamilies: (1) the Hyperoodontinae, which contains *Hyperoodon* and *Mesoplodon*; (2) the Ziphiinae, which contains *Ziphius, Berardius, Tasmacetus,* and the fossil genera *Choneziphius, Ziphirostrum, Cetorhynchus,* and *Ninoziphius*; and (3) the Squaloziphiinae, which currently contains only *Squaloziphius.* Since de Muizon's work, several more genera of fossil ziphiids have been found, but it is unclear how these genera relate to the modern beaked whales. Of course, not everyone agrees with de Muizon's classification, but it is a reasonable starting point for conversation.

The living mesoplodonts are quite similar and are differentiated by the shape and position of the tusks, believed to be used primarily "as weapons in aggressive encounters with other males . . . females are effectively toothless" (Dalebout, Steel, and Baker, 2008). "Examples of weapons of male-male competition," wrote Dalebout et al., "include the antlers and horns of ungulates; the claws of amphipods, and the horns of rhinoceros and dung beetles . . .

secondary sexual characteristics can also fulfill dual functions, as weapons of male-male competition and as cues for female choice or species recognition. Such traits often arise through male-male competition and serve as honest signals to other males regarding fighting ability or dominance."

If the mesoplodonts are so similar, why are there so many tooth shapes? Dalebout and colleagues suggested four hypotheses to explain the development of the different tusk shapes in the various species:

- Those species with apical, forward-facing tusks were more primitive (i.e., earlier), whereas those with backward-facing tusks set back from the apex are more derived (i.e., later developments).
- Speciation was driven by population isolation in deep-sea canyons, isolating one population from another and allowing for the development of different tusk morphology.
- Sexual selection may be the driver of speciation, with differing tusk shapes allowing different species to occur in the same ocean basin.
- Variation in tusk shape and position could function as species recognition cues and function as a precopulatory isolating mechanism.

The authors write that only sexual selection (#3) and species recognition (#4) "seem to warrant further consideration" as shapers of tusk morphology, because males and females would be able to identify conspecifics visually and probably also identify each other by echolocation. They concluded:

> We have tested predictions from four hypotheses regarding the evolution and diversification of a speciose genus of whales using a robust, highly resolved molecular phylogeny. The patterns observed in three of the four sister-species pairs identified strongly suggest that sexual selection on weaponry in the form of male tusks, together perhaps with selection for species recognition cues, has played an important role in this unique radiation. To our knowledge, this is the first time that sexual selection has been explicitly implicated in the radiation of a mammalian group outside terrestrial ungulates.

The tusks of beaked whales are a largely ornamental trait that became a driver in species separation. Because they erupt only in adult males, the protruding tusks help the females to identify sexually mature males within their species. The scars that result from male-male fighting are probably an additional means of signaling breeding readiness and dominance, directly related to the use of the tusks. This, of course, is Colin MacLeod's theory, as elucidated in his 1998 paper "Intraspecific Scarring in Odontocete Cetaceans: An Indicator of Male 'Quality' in Aggressive Social Interactions," where he wrote,

> The level of visible (i.e. white or unpigmented) scarring on cetaceans varies greatly between species, particularly for intraspecific scarring in odontocete cetaceans. In some species, unpigmented intraspecific scars may act as an indicator of male 'quality' during aggressive social interactions. Evidence to support this hypothesis was found in 18 species of odontocete cetacean. These were the narwhal (*Monodon monoceros*), the sperm whale (*Physeter macrocephalus*), the Risso's dolphin (*Grampus griseus*) and the family Ziphiidae (with the exception of *Mesoplodon ginkgodens*). The evolution of such signaling is related to the fact that teeth are not required for feeding on certain diets, primarily cephalopod-based diets, and as a result the number of teeth has been reduced. However, some teeth have been retained, and selected, as weapons for male-male competition. This has resulted in an increase in the level of intraspecific scarring and the greater need for a signal of "quality" to avoid costly and dangerous fights. As intraspecific scarring became this signal, the re-pigmentation rate of scars was reduced, leading to all scars remaining permanently unpigmented in these species.

Beaked Whales versus the Navy

> Whether owing to the almost omniscient look-outs at the mast-heads of the whale-ships, now penetrating even through Behring's straits, and into the remotest secret drawers and lockers of the world; and the thousand harpoons and lances darted along all continental coasts; the moot point is, whether leviathan can long endure so wide a chase, and so remorseless a havoc; whether he must not at last be exterminated from the waters, and the last whale, like the last man, smoke his pipe, and then himself evaporate in the final puff.
>
> Herman Melville, *Moby-Dick; or the Whale*, 1851

At one time or another over the past thousand years, nearly every species of great whale has been hunted to the brink of extinction. The failure of sixteenth-century European explorers to find the Northwest or Northeast Passages turned out very badly for the bowhead whales they found instead:

reports to the ship owners of Holland and Britain resulted in the rise of the European whaling industry, where the baleen whales of the Western Arctic—also known as Greenland or polar whales—were nearly exterminated, to such a degree that the industry itself had to shut down. Right whales, usually the first to be hunted when a new region was settled—think of Cape Cod or Tasmania—were among the first casualties, and although some populations have rebounded, northern and southern right whales are still considered endangered. The rorquals—the blue, fin, sei, Bryde's, and minke whales—were slaughtered in such enormous numbers in the Antarctic that hunting had to be suspended to protect the last survivors. In the Antarctic, the scene of the whaling industry's revival after the collapse of the nineteenth-century Yankee sperm whale fishery, blue whales were the first targeted by Norwegian and British whaling fleets and in 1931, 29,000 blue whales were killed in a single season. The gray whales of the Eastern Pacific were declared extinct in the late nineteenth century, but they "recovered" during the twentieth. Humpbacks, the singing whales that are the darlings of whale watchers everywhere, were depleted throughout their range, and now their populations are mostly relicts of their former abundance. After pursuit by Nantucket and New Bedford whalers in the nineteenth century, and by Norwegian, British, Japanese, and Soviet whalers in the twentieth, the worldwide population of sperm whales is but a shadow of its previous numbers.

The overall decline in large whale populations, along with passionate objections to whaling per se, were the driving forces behind the passage of the International Whaling Commission's 1983 moratorium on commercial whaling. The whaling nations that voted against the moratorium, and who wanted to continue the slaughter indefinitely, paid no attention to the beaked whales, because with the exception of Baird's beaked whale and maybe the northern bottlenose, they were too small, too rare, too hard to find, and simply not worth the effort. Besides, we knew so little about their numbers (or even where some of them lived) that they came in, as it were, under the radar, free to live their deepwater, pelagic lives outside the threat of grenade harpoons. But in recent years, in our never-ending search for novel and more efficient technologies, we have found new ways to threaten all kinds of whales, especially the beaked ones.

We still know very little about where they live—it might be said that we know more about where they die—but a 2013 study by Jeffrey Moore and Jay Barlow on the "Declining Abundance of Beaked Whales in the California Current Large Ecosystem," opens with these words:

> While little is known about beaked whale ecology, and in spite of their cryptic existence, there is nevertheless a long list of documented human impacts to beaked whales. [They] are hunted, entangled unintentionally in fishing nets, affected behaviorally and physiologically by naval active sonar (often with lethal effects), disturbed to an unknown extent by other ocean noise sources such as from large commercial vessels or oil and gas seismic surveys, susceptible to health effects of ingesting plastic debris, and potentially vulnerable to deepwater ecosystem changes driven by climate-related oceanographic forcing or other human impacts such as demersal fishing.

A gillnet is a mesh net deployed to catch fish that swim into it. The nets can be made to hang suspended between the bottom and the surface, float on the surface, or be set in place with anchors. Drift gillnets, usually composed of multifilament mesh, are allowed to drift with the current. The drift gillnet fisheries of the California coast target swordfish, thresher sharks, and shortfin makos, but the nets fish indiscriminately, entangling (and drowning) anything that blunders into them, including dolphins, seals and sea lions, sea turtles, seabirds, and beaked whales. According to a 2003 study by Jay Barlow and Grant Cameron, acoustic deterrent devices (pingers) were first applied to the nets in 1996, in an effort to keep animals (other than the target fish species) away from the nets, and while the pingers had little effect on dolphins and pinnipeds, they were 100% successful in eliminating beaked whale bycatch: "From 1990 to 1995, biologists observed 33 beaked whales entangled in 3,303 fishing sets," wrote Caretta, Barlow, and Enriquez in 2008, "Since pinger use began in 1996, there have been no beaked whale entanglements in 4,381 observed sets through 2006." It is possible that the drop in beaked whale abundance in California waters (Moore and Barlow 2013) might be the reason for fewer entanglements, but beaked whales' known sensitivity to man-made noises might also explain why they stayed away from the pingers.

Cetaceans have been stranding for much longer than there were men to find them on the beach. We do not understand these apparently self-destructive activities of the whales, but they have variously been attributed to illness, fear of predators, unusual tides or currents, failure of the navigational system, migraine headaches, pregnancy, brain tumors, suicide wishes, protests against whaling, infectious disease, biotoxins, parasites, oil spills, and even a desire to return to the origins of their land-based ancestors. Perhaps some strandings can be attributed to debilitating illness, when the animal is so stressed that it takes itself into shallow water to keep its blowhole above water, so that whatever else happens to it, at least it will not drown. Whales in water too shallow to swim in almost invariably end up on the beach, where they bake in the sun. Sometimes cetaceans come ashore and die in substantial numbers.

There is a (one-man) school of thought that holds underwater earthquakes (*seaquakes* or *T-waves*) responsible for whale strandings. David Williams, a retired ship captain from North Carolina, claims that his 2016 report "Why Whales Strand" includes the first use of the words *deaf, deafness, auditory trauma, hearing injury,* and *sinus barotraumas* in a discussion of whale strandings. Williams quotes a 1959 study by Allan Milne comparing the noise level of undersea earthquakes to that of a nuclear explosion: "Travel time measurements and signal spectra indicate that the nuclear explosions [that] originated within Eniwetok Atoll . . . were similar in nature to that of the earthquake *T*-phase, but the duration of the signals from the nuclear explosion was considerably less." Certainly an underwater explosion of similar magnitude would injure whales anywhere in the vicinity, but injured whales in the middle of the ocean will probably die, sink, or be eaten by sharks before they wash ashore. Williams' hypothesis indicates that odontocetes mass beach themselves about 26 days after exposure to an injurious seaquake. Some injured pods recover; those that do not swim downstream with the flow of the surface current about 4,000 km (2,200 nautical miles) before they are washed ashore onto the beach by a strong shoreward flow. Undersea earthquakes occur most frequently along the mid-ocean ridge system: at the epicenter of 85 percent of all the seismic activity on our planet, the species of whales and dolphins repeatedly mass beaching dive and feed in the most earthquake-prone area on Earth. The species that do not reside in earthquake hotspots do not mass beach.

Among dolphins, the most notorious mass-stranders are false killer whales (*Pseudorca crassidens*)—a single stranding of 835 animals was reported from Mar del Plata, Argentina, in 1946—and the pilot whales, whose enforced strandings have formed the basis for a harvesting industry in Scotland, Iceland, and the Faroe Islands. Almost all toothed cetaceans have been known to beach themselves individually or in groups, and while there have been many hypotheses, no clear explanation has yet emerged for this puzzling phenomenon. Aristotle wrote, "It is not known for what reason they run themselves aground on dry land; at all events, it is said that they do so at times, and for no obvious reason," and this is as true now as it was in the fourth century BC, except, as we shall see, for those recent strandings that can unequivocally be blamed on people.

The North Sea coast of Holland would appear to be one of those places (noteworthy others are in New Zealand and Cape Cod) where whales strand with some degree of regularity. From 1531 to around 1690, some nineteen whales of assorted species beached themselves on these shingled coasts. Most of them seem to have been sperm whales, and with their huge heads, their jawfuls of ivory teeth, and—in what appear to be a majority of the cases—their male genitalia prominently exposed, the dead whales must have been a wonder of wonders to the Dutch persons who came to view these beached leviathans. (It would be another half-century before the whalers of Rotterdam and Delft would head for the icy seas of Spitsbergen, where they would hunt a totally different creature, the Greenland right whale.)

Although the northwestern quadrant of the North Sea is somewhat unfamiliar territory for sperm whales, the shores of Denmark, Germany, the Netherlands, and Belgium have experienced stranded cachalots for at least four centuries. Indeed, these events are so common that in 1995, a symposium was convened at Koksijde, Belgium (Jacques and Lambertsen 1997), to discuss sperm whale strandings in the part of the southeastern North Sea known as the Wadden Sea (*Waddenzee* in Dutch; *Wattenmeer* in German), an intertidal zone between the coast of northwestern continental Europe and the Frisian Islands, forming a shallow, semi-enclosed body of water—the perfect environment for sperm whale strandings. The events are in no way related to beaked whales, but probably the most remarkable beaked whale stranding of all also took place in the Netherlands; in 1927, a single Gray's beaked whale (*Mesoplodon grayi*) stranded at Kijkduin, 7,000 miles from its archetypical New Zealand habitat.

Strandings cannot be predicted, so coming across a carcass on the beach or a whale floundering in the shallows has to be seen as serendipitous. Mother and calf Gray's beaked whales swam around a New Zealand harbor for five days—an obvious prelude to a stranding—and then swam off. Skeletons, whole or partial, are sometimes found by beachcombers looking for something else or dog-walkers out for a stroll on the beach. Rarely—as with almost all beaked whale discoveries—skulls are found in museum collections, mislabeled or not labeled at all because the curators could not figure out what they were. Hurricane Sandy, the enormously destructive "superstorm" that roared ashore in the eastern United States in late October 2012, produced perhaps only one valuable result: a 9-foot-long beaked whale skeleton was uncovered by beach erosion on New Smyrna Beach on the Atlantic coast of Florida, south of Daytona. Although the skull was found largely intact, Volusia County officials are still looking for the lower jaw, which is required to identify the species.

Without strandings, whatever their fateful inspiration, the beaked whales would be even more poorly known than they are now. As beaked whale expert Robert Pitman wrote in his 2009 discussion of the mesoplodonts in the *Encyclopedia of Marine Mammals*,

> Mesoplodont whales are found in nearly all of the deeper oceanic waters of the world, and they are by far the most speciose genus of marine mammal. But despite this, they remain among the least known large animals on the planet. Several species have never been identified alive in the wild, and at least one species is known only from skeletal remains . . . Because of their shy nature, oceanic habitat, an apparent rarity, very little is known about the biology of mesoplodonts, and nearly everything that is known has come from the examination of stranded animals.

Most beaked whales are studied after life has ended, as stranded animals. Biologists who would study stranded whales consider themselves fortunate when they have a carcass to examine, but for every pregnant, sick, injured, or disoriented beaked whale that ends up in water shallow enough to strand, there must be an uncountable number that die in deep water and are therefore never seen by researchers. Whatever the "reasons" for strandings, they have been a boon to whale researchers, enabling them to identify old and new species, but at the same time, they are an alarm call to those who recognize that sonar testing may be implicated in the death of the stranded whales.

Ironically, the techniques that cetaceans use to navigate, find food, and find each other are the very things that are killing them. Echolocation is actually a form of sonar, where dolphins, sperm whales, and beaked whales generate high-pitched sound pulses or clicks from somewhere in their heads and "read" the returning echoes to identify what's in front of them, what it's made of, and how far way it is. Where the Navy reads the returning echoes on a screen, dolphins (and perhaps beaked whales) sense the returning sound vibrations by feeling the pulses in their jaws. Every underwater object or animal sends different echoes, which the dolphins can differentiate. You can't see very far through water no matter how good your eyes are, but water is an excellent sound transmitter, transmitting sound five times faster than air. Whales and dolphins have capitalized on the properties of water to develop an entire communications system based on sound, and although we're not sure exactly how it works, many cetaceans use sound not only to locate their prey, but maybe even to disable it.

All whales and dolphins depend upon sound as a significant component of their communications repertoire: dolphins click and whistle, humpbacks and bowheads sing, killer whales scream, and sperm whales click, moan, groan, creak, bang, and clang, all these noises employed as inter- (or intra-) species communications of one sort or another. Sound for cetaceans is the analog to vision in terrestrial vertebrates; light rays enter our eyes and the image is processed by the brain; sounds enter the hearing mechanisms of whales and dolphins and are processed by their brains. It stands to reason, therefore, that the introduction of ear-splitting non-cetacean sounds into the whales' environment will be confusing, disturbing, or even detrimental. (Think of a person trying to find his way around a room with a battery of strobe lights flashing on and off.) When Australian underwater filmmaker Ben Cropp was filming humpbacks off Port Douglas, northern Queensland, he recorded a mother and calf taking refuge on the side of his boat away from a military vessel and said that this was not the first time he had observed this behavior.

In the introduction to her 2007 article on the effects of noise on marine mammals, Linda Weigert wrote,

> Cetaceans are highly vocal and dependent on sound for almost all aspects of their lives, e.g., food-finding, reproduction, communication, detection of predators/hazards, and navigation . . . Observed effects of noise on marine mammals include: changes in vocalizations, respiration–swimming speed, diving and foraging behavior, displacement, avoidance, shifts in migration path, stress, hearing damage, and strandings . . . Acoustically-induced strandings may displace a local beaked whale population (for an extended period if not permanently) or even possibly eliminate most of its members. As beaked whales seem to be found in small, possibly genetically isolated, resident populations, even a transient and localized acoustic impact could have prolonged population consequences.

Shipping is most likely the main overall source of man-made noise in the marine environment, but non-shipping vessels, such as pleasure boats and whale-watching boats, contribute to the ambient noise level in the vicinity of whales and dolphins, and in some places, such as Shark Bay in Western Australia, the noise of the dolphin-watching boats caused the dolphins to seek other premises. Other problems are seismic exploration, which consists of setting off underwater explosions and timing the returning sounds to ascertain the geological composition of the bottom; pile driving and construction; underwater explosives and blasting; offshore oil exploration and production, shattering icebergs; and of course, activities whose primary function is to *make* noise while being entertained, such as riding Jet Skis. However, the use of military sonar appears to be the main villain.

The Allies introduced sonar (an acronym for SOund NAvigation and Ranging) during World War II to track German U-boats in the Atlantic. Sonar locates submerged objects by sending sound waves through the water and reading the range, bearing, and nature of the target by the returning echoes. Surface and submarine vessels use low-frequency active sonar (LFAS) as a navigational aid. Military ships use mid-frequency sonar (MFS) by firing bursts of sound through the water and listening for an echo off a ship's hull. Blasted from subsurface loudspeakers and aimed not unlike a powerful searchlight on land, sonar powerfully (and noisily) scans the surrounding depths. Some mid-frequency sonar can put out more than 235 decibels, as loud as the launch of a Saturn V rocket. Such blasts can affect the sensitive hearing apparatus of whales and dolphins in the vicinity, injuring them and causing them to beach themselves or otherwise behave strangely.

When a human scuba diver descends to significant depths, nitrogen, which constitutes four-fifths of every breath, is forced into solution under the ambient pressure. As the pressure falls—that is, as the diver rises to the surface—the nitrogen again turns into bubbles of gas, and if not allowed to dissolve, can obstruct blood vessels. This is called "the bends." If the bubbles obstruct the blood flow to the heart or the brain, they can cause death. Divers who have been at depths for any amount of time have to decompress, either by waiting at designated stations for a prescribed amount of time or, in an emergency, by entering a decompression chamber. For the most part, deep-diving animals, such as certain whales, seals, and penguins, do not get the bends because, while human divers take more than one breath, the other animals submerge after taking only one breath, and the air is compressed into "safe" regions of the respiratory pathway, such as the air sinuses at the base of the skull. Rather than possessing a particularly large lung capacity, the deepest diving whales, such as sperm whales and beaked whales, have relatively small lungs, and at depths of 300 feet or more, their lungs collapse completely, forcing all the air into the rigid respiratory dead spaces.

On May 12, 1996, twelve Cuvier's beaked whales (*Ziphius cavirostris*) stranded and died along the shore of Kyparissiakos Gulf in Greece. Necropsies revealed no evident abnormalities or wounds, but it was learned that LFAS "sound-detecting system trials" had been performed by the NATO research vessel *Alliance* the previous day. Volunteers managed to push five of the whales back into the ocean, but seven died. "Although pure coincidence cannot be ruled out," wrote Frantzis in 1998, "it seems improbable that [the strandings and the sonar tests] were independent. Little is known about whales' reactions to LFAS; to obtain definitive answers, more information needs to be gathered. But unfortunately, most of the data about the use of LFAS are subject to military secrecy."

From March 15 to March 20, 2000, fifteen beaked whales, a spotted dolphin (*Stenella frontalis*), and two minke whales (*Balaenoptera acutorostrata*) stranded or became trapped in the shallows of the Bahamian islands of Abaco, North Eleuthera, and Grand Bahama. During this period, U.S. Navy warships were conducting sonar tests and maneuvers in the area. Ken Balcomb, whose Center for Whale Research in Friday Harbor, Washington, has been studying killer whales since 1976, initiated a project in 1991 to photographically document the beaked whale populations of the Bahamas. Thus he was on Abaco Island in March, 2000, when the cetaceans stranded. Five Cuvier's beaked whales, the spotted dolphin, and two other beaked whales (one *M. densirostris* and one *M. europaeus*) died. In 2001, with Diane Claridge, Balcomb wrote a detailed article for the *Bahamas Journal of Science*, which he called "A Mass Stranding of Cetaceans Caused by Naval Sonar in the Bahamas."

In a plea for more research on the risk of acoustic impact on beaked whale populations, ten cetologists (Taylor et al. 2004) said,

> The worldwide increase in uses of high-intensity underwater sound raises serious conservation and management concerns for this suite of species . . . Collaborative research between acousticians and biologists is needed to assess the magnitude of the problem. Such research should include mapping of current and planned highintensity acoustic activities, mapping of beaked whale densities and identifying areas where no data are available, estimating beaked whale densities and population structures particularly in areas of high acoustic activity, investigating the mechanism of harm to beaked whales from high-intensity acoustics, developing improved beaked whale detection methods, and estimating the probability of detecting lethal effects on beaked whales.

The hearing of cetaceans has to be much more sensitive than that of land mammals (but probably not of bats), because they depend on sound to earn a living. Injury occurs when there is a permanent threshold shift that results in loss of hearing, likely to occur "before non-hearing pressure injuries (barotraumas) caused by sound." Balcomb and Claridge wrote, "As many readers know from personal experience, pressure trauma in the airspaces of the middle ear and eustachian tubes can be extremely painful, no matter how it is caused (disease, diving, air travel, etc.). Intense sound pressure can also be painful and injurious to humans and animals, and can result in either auditory or non-auditory barotraumas." The pain you may have felt in your ear while diving or coming in for a landing in an airplane is only a faint shadow of the trauma inflicted upon the beaked whales by sonar's sonic blasts, so intense that it ruptures the airspaces and may even kill the whales. "Post mortem examinations of the beaked whales that stranded in the Bahamas revealed bleeding around their eyes, ears, and brains, which is consistent with acoustic trauma," wrote Taryn Kiekow in her 2009 discussion of beaked whales and sonar.

The etiology of beaked whale deaths is not known. In conjunction with sonar testing, some beaked whales (mostly *Ziphius*) "run themselves aground on dry land," but even when scientists necropsy the carcasses, the cause of death is not immediately evident. We know that cetaceans rely on sound more than sight for communication, but we're not sure how they hear. We only know that they don't hear the way we do. Although all odontocetes have external ears, they are usually no more than a pinhole located behind the eye, and the meatus (the auditory canal) and the eardrum no longer function in the hearing process. The dolphin hears through the lower jaw, the process McCormick and colleagues first described in 1970. In *The Porpoise Watcher* (1974), Norris devotes an entire chapter to "The Jaw-Hearing Porpoise" (Norris preferred to call dolphins "porpoises") and discusses his finding of a porpoise jawbone on a remote beach in Baja California:

> What peculiar jaws they were! Most land animals have stout jaws adjusted to the forces of chewing or tearing, or of pulling up grass or the leaves from trees. One can see strong flanges where the jaw muscles attach. Nothing like this seemed to be represented in the porpoise jaw. In fact, it was so thin towards its rear end that I could see sunlight shining through the translucent bone, which at its center was less than a millimeter thick . . . Nothing in nature as bizarre as this is created unless the forces of survival hold out value for it. Why, I wondered, did a porpoise find it useful to have a jaw this thin and delicate as the finest porcelain, and a nerve and blood vessel canal that consumed the whole rear end of the jaw?

Norris identified the actual sound receptors as pockets of fat in each side of the mandible, which transmit the incoming sound to the bulla (ear bone), "thence to this tiny bone strut to the inner ear and brain." Continued experimentation over the next five years proved that Norris was correct, for in 1979 he wrote, "It seems clear now that the environmental sounds enter the odontocete mandible by passing through overlying blubber through what I have termed the 'acoustic window,' penetrating the very thin 'pan bone' of the rear mandible, enter the mandibular fat body and are transmitted to the middle ear."

It has been easy enough to test the hearing of bottlenose dolphins, but beaked whales, rarely seen at sea and never maintained in captivity, are another story. We're not altogether sure what they hear *with*. Balcomb wrote that

> in delphinids the pterygoid sinuses are laterally and ventrally encapsulated with a thin laminate of bone and they are much smaller than the ziphiids. We propose that the relatively enormous open-sided airspace that has evolved in ziphiids could serve not only as an air reservoir for the middle ear, but also as an acoustic mirror that would increase the sensitivity of 'jaw-hearing' over that of delphinids and other cetaceans that lack it. Such hearing advantage at depth would support the surmise that these whales evolved as sensitive listeners more than active echolocators of their primary prey (squid)."

Particularly "sensitive listeners" would also be more susceptible to loud sonar blasts under water. After weeks of denial, the U.S. Navy admitted that they were holding naval maneuvers in the Providence Channels, exactly where the whales had stranded, and that the Navy was responsible for the whale deaths.

There must have been cases where the whales died and did not come ashore, so the records have to be viewed as incomplete, but for reasons not clearly understood, Cuvier's beaked whales are the most frequent victims of sonar testing. (They are among the most widely distributed and the deepest diving of all cetaceans.) On the Portuguese island of Madeira, 120 miles north of the Canary Islands in the North Atlantic, three Cuvier's beaked whales mass-stranded between May 10 and May 14, 2000. A fourth animal was reported floating in the water by a fisherman, but it did not come ashore. A NATO naval exercise off Portugal started just one day prior to the stranding. The head of one beaked whale was in adequate condition to be examined, and it was found to have blood in and around the eyes, ears, and brain. This animal and another that were also examined on site were also found to have lung hemorrhaging.

On September, 24, 2000, two dead Cuvier's beaked whales were found on the beach of Mexico's Isla San José, in the southern Gulf of California. Investigators learned that the Columbia University research vessel *Maurice Ewing* had been conducting seismic surveys in the area, which consisted of "bouncing sound pulses produced by blasts of compressed air off the gulf's floor to map the margins of the continental plate" (Malakoff 2002). Columbia University scientists denied any connection between the dead whales and the seismic testing—they claimed the ship was 30 miles from the stranding site—but they temporarily suspended operations anyway. "Remarkably," wrote Samuel Malakoff in *Science*, "the Mexican incident occurred on the same day that more than a dozen beaked whales stranded off the Canary Islands, in the eastern Atlantic, following naval exercises conducted by U.S. and Spanish vessels."

From September 24 to 27, 2002, fourteen beaked whales stranded on the southeast coast of the island of Fuerteventura and the northeast coast of Lanzarote in the Canary Islands. At the time of the strandings, a NATO multinational naval exercise known as *Neo Tapon* was being conducted off the Canaries, where "at least one aircraft carrier, 50 surface vessels, 6 submarines and 30 aircraft were participating in acoustic exercises," so it can reasonably be assumed that tactical hull-mounted sonar systems were in operation (D'Spain, D'Amico, and Fromm 2006). Scientists were able to recover eleven carcasses: nine *Ziphius cavirostris*, one *M. densirostris* (previously found together in the waters of Hawaii and in the stranding in the Bahamas), and one *Mesoplodon europaeus*, known as Gervais' beaked whale. Necropsies showed that the dead whales had symptoms similar to those of decompression sickness (the bends), which is uncommon, but not unheard of, in whales and dolphins.

In their survey, Moore and Barlow (2013) found that "abundance of *Z. cavirostris* and *Mesoplodon* spp. in the California Current has likely declined over the 18-year study period of 1991 to 2008. The evidence is particularly strong for *Mesoplodon*, based on our analysis of survey-cruise data as well as the strandings record." Among the major causes for this decline, they list mortality from fishing (bycatch); ecosystem changes; and Navy sonar and other anthropogenic ecosystem noise. "*Ziphius* and *Mesoplodon* are the two beaked whale genera known to suffer impacts from naval sonar activities. They exhibit strong behavioral responses to certain types of active sonar, resulting in altered movements and space use for prolonged periods after exposure. In more extreme cases there can be physiological consequences leading to death or stranding."

Beaked whales are among the deepest diving species, so sonar blasts might cause them to change their normal diving pattern and come to the surface faster, which would cause nitrogen bubbles to form in the tissues, internal injury, and possibly death. As Sascha Hooker, Robin Baird, and Andreas Fahlman wrote in the journal *Respiratory Physiology and Neurobiology* in 2009, "Beaked whales dive deeply more frequently than most other cetacean species and this behavior has been suggested to to result in tissue N_2 supersaturation. It is possible that these mammals live continuously with elevated levels of N_2 which could render them prone to decompression sickness (DCS) if they altered their behavior. Suggestions include disturbance caused by an acoustic signal that could affect the normal diving behavior, e.g., increased or decreased suface interval, modified ascent rate or dive duration, leading to increased supersaturation, thereby increasing DCS risk." Indeed, when scientists necropsied the beaked whales that stranded four hours after mid-frequency sonar activity in the Canaries, they found that the animals "showed severe, diffuse vascular congestion and marked, disseminated microvascular haemorrhages associated with widespread fat emboli within vital organs. Intravascular bubbles were present in several organs, although definitive evidence of gas embolism *in vivo* is difficult to determine after death." In other words, the sonar probably killed them.

Introducing "A Note on the Unprecedented Strandings of 56 Deep-diving Whales along the UK and Irish Coast" (2010), Dolman, Pinn, Reid, and eleven other authors wrote: "Between 21 January and 27 July 2008, there were unprecedented strandings of eighteen Cuvier's beaked whales, four Sowerby's beaked whales, five unidentified beaked whales, and twenty-nine long-finned pilot whales in Scotland, Ireland and Wales. Most were assumed to be dead upon stranding. In addition to those in the UK and Ireland, there was a mother and dependant calf pair of Sowerby's beaked whales stranded live at Calais, France on 19 January 2008. The mother died and the calf was lost alive back at sea." The whales, all of which are believed to have stranded individually around mid-January, were found in various stages of decomposition (some had been reduced to skeletons), and the cause of death (or stranding) could not be determined. In fact, the condition of the carcasses did not necessarily indicate that the whales had been driven ashore to die. As the authors wrote, "Care should be taken with the interpretation of strandings data as they might not be representative for determining events that happen at sea. Clearly, whales that die at sea can sink, refloat, drift, or be eaten by predators and may never strand. A further question remains about the number of animals that died, as additional carcasses may have sunk at sea and never stranded." Although the death of so many cetaceans in such a short time period strongly suggests the use of mid-frequency active sonar, the British Ministry of Defense (MoD) denied any naval activity in the area in January 2008. The authors also considered seismic surveys, fisheries interactions, disease, and seaquakes, and tentatively concluded that "it remains a possibility that a currently unidentified anthropogenic or natural factor may have been contributory to these mortalities." (It is also possible that the death of these cetaceans occurred at a time when the MoD *was* conducting naval activities.)

The Spanish government imposed a moratorium on naval exercises in the waters of the Canary Islands in 2004, and since then there have been no recorded beaked whale strandings (Fernández, Arbelo, and Martín 2013). Although far from conclusive, the nine-year moratorium would, at least circumstantially, lean toward the implication of sonar in beaked whale strandings. But the way in which sonar actually impacts the whales within range has not been clearly identified. "Initially," wrote Allen, Schanze, et al. in 2013, "it was hypothesized that the sonar caused direct physical damage to the whales, due to the presence of gas bubble lesions and subarachnoid hemorrhages observed in stranded animals, and the potential for intense sound energy to cause bubbles to grow in supersaturated tissues," as suggested by Hooker, Baird, and Fahlman, above. Allen, Schanze, et al. continue: "More recent hypotheses have focused on the possibility that sonar initiates a chain of events that lead to strandings but starts with a purely behavioral reaction. Beaked whales live in deep waters, so they must show a strong avoidance reaction to swim from their normal habitat onto the beach." Besides sonar, what else might stimulate such a "strong avoidance reaction"? How about the presence of noisy predators?

With a range that includes all the world's oceans, including tropical and high polar latitudes, killer whales are the most widespread of all cetaceans, and are every other whale's worst enemy. They hunt in packs and have been known to attack just about every large aquatic creature, including blue whales, gray whales, humpbacks, sperm whales, dolphins, porpoises, seals, sea lions, sharks, fishes, squid, and an occasional sea bird. Their attacks usually take place at or near the surface, so killer whales are not designed to—nor do they need to—dive particularly deeply. Despite its common name, *Orcinus orca* is the largest of the dolphins, with large males reaching a length of 30 feet and a weight of 6 tons or more. Like all dolphins, they are accomplished echolocators and are fast enough to catch almost anything that swims. They hunt in packs (usually composed of family groups), and they can tear their prey to pieces with their powerful jaws and teeth. Some whale species have been observed to throw themselves onto the beach when being chased by killer whales (Ford and Reeves 2008), and it is likely that the sounds of hunting killer whales would terrorize any cetacean that hears them coming. In the varied sound repertoire of killer whales, clicks, cries, and whistles are used for echolocation and communication, but they can also produce loud screams, which they use to herd and panic prey animals, from herring to humpbacks.

As far as we know, there is no evidence of a killer whale attack on a beaked whale, but given the wide range of both, occasional interactions are likely. Off Andros Island in the Bahamas, scientists played recordings of the vocalizations of "marine mammal-eating killer whales" to a tagged adult female Blainville's beaked whale to observe her reactions to the sounds. The beaked whale responded with "a prolonged and directed avoidance reaction"; that is, she swam directly away from the killer whale sounds and kept going. Naval mid-frequency sonar elements were also played, but the whale's response was not as pronounced or prolonged. If the fake killer whale had been programmed to follow the fleeing beaked whale close to shore, it is not unreasonable to assume that "the prolonged directed avoidance response observed here suggests a behavioral response that could pose a risk for stranding" (Allen, Schanze, Solow, and Tyack, 2013). Obviously, broadcasting louder and louder killer whale sounds will cause a change in the behavior of a beaked whale and could, under the right circumstances, cause the panicked beaked whale to beach itself, but this does not absolve naval sonar from responsibility for the strandings and deaths of many cetaceans.

In a 2011 discussion called "No Shallow Talk," Natacha Aguilar de Soto and several colleagues pointed out an inherent problem: "Communicating animals must balance fitness benefits against the costs of signaling, such as increased predation risk. Cetaceans communicating mainly with sound and near-surface vocalizations can place signalers at risk from shallow-diving top predators with acute hearing, such as killer whales." So most of the sounds made by Blainville's beaked whales are designed for locating (and maybe debilitating) prey items during deep foraging dives, but tagged whales off El Hierro in the Canaries were recorded making "two novel types of sound apparently dedicated to social communication: whistles and a rapid series of clicks that we named rasps." These whales live and dive in small groups, so they need to communicate in order to coordinate their activities and maintain social ties. Although this coordination would seem to require some shallow or surface communication, none has ever been recorded, and these whales are totally silent at the surface, whistling and rasping at depths of 3,000 feet. In other words, if you make too much noise in shallow waters, killer whales will find you.

It now appears that the threat of killer whale predation has played a role in the evolutionary development of certain cetaceans. Just as beaked whales restrict their echolocating pulses to depths out of the killer whale's range, certain porpoises have limited their click energy to a narrow band under 100 kHz—the upper limit of killer whale hearing. Harbor porpoises (*Phocoena phocoena*) are small, slow-swimming animals that produce high-frequency clicks for echolocation (and probably intraspecies communication) that are out of the killer whales' hearing range. The authors of this 2013 study (Kyhn, Tougaard, et al.) suggest that the otherwise defenseless porpoises developed an effective sonar system that concurrently minimizes the risk of killer whale predation.

A 2011 study ("Beaked Whales Respond to Simulated and Actual Navy Sonar" (Tyack, Zimmer, et al.) showed that at Andros Island, where Blainville's beaked whales had been known to strand in the past, *playback* of sonar sounds caused the whales to stop echolocating and make "unusually long and slow ascents from their foraging dives." And now it has been shown that sonar—real or simulated—is not the only thing that affects them. A 2012 study, "Vessel Noise Affects Beaked Whale Behavior" (Pirotta, Milor, et al.), showed that "other forms of acoustic distubance (e.g. shipping noise) may disrupt behavior." The researchers tagged one adult male Blainville's beaked whale in Tongue of the Ocean, a deepwater basin to the east of Andros, and monitored his reponses to the various sounds. In the presence of simulated military sonar, killer whale sounds, and band-limited noise, the whale not only stopped echolocating, he moved some distance from the sound source. "There is no evidence," wrote the authors, "that beaked whales have stranded during periods when naval mid-frequency active sonar (MFA) is being used. However, our data show that beaked whales move tens of kilometers away from sonar excercises there. The avoidance reponses reported here reduce exposure to sonar, but if beaked whales move out of their normal deep-water habitat into shallow water, this could increase their risk of stranding." Whales obviously cannot strand in deep water, so driving them into the shallows exponentially increases their chances of coming ashore.

Not only beaked whales are affected by military sounds. During a week of British naval war games in June 2008, some sixty common dolphins (*Delphinus delphis*) swam into Falmouth Bay, near the far western end of Cornwall, open to the waters of the western English Channel. At that time, a British "live fire" task force, including twenty Royal Navy ships, helicopters, and submarines (including the nuclear sub HMS *Torbay*) and eleven foreign vessels, was deployed to the area. On June 7, the dolphins swam into Porth Creek, a dead-ending arm of the Bay, and by the 9th, they were beginning to come onto the beach. Some were rescued and herded back to sea, but twenty-six died, in what became the largest mass-stranding event (MSE) in British history. In a comprehensive report issued five years later (2013), Jepson et al. observed that all the dead dolphins were immature, and five were calves. None showed signs of infection or any other incapacitating condition, so it was assumed that the noise from the ships and helicopters drove the animals into the shallows where, in a panic, they threw themselves onto the strand. "International naval excercises," wrote Jepson et al., "did occur in close proximity to the MSE with the most intense part of the excercises (including mid-frequency sonar) occuring four days before the MSE and resuming with helicopter excecises on the morning of the MSE. The MSE may therefore have been a 'two-stage process,' where a group of normally pelagic dolphins entered Falmouth Bay and after 3–4 days in/around the Bay, the second acoustic disturbance occurred, causing them to strand *en masse*."

Melon-headed whales (*Peponocephala electra*) are large black dolphins that inhabit deep tropical and temperate waters worldwide. On July 3, 2004, a group of around two hundred were seen in Hanalei Bay off the Hawaiian island of Kauai, swimming in abnormally tight circles a hundred feet from shore. They milled around for most of the day and night in an agitated manner, tail slapping and vocalizing. An NOAA (National Oceanic and Atmospheric Administration) rescue team herded the animals back out to sea, but one beached itself and died. A U.S.-Japanese naval training exercise was being held in the vicinity, and even though the Navy originally claimed that they had used their sonar *after* the whales' erratic behavior, it was later revealed that they had been testing the sonar the previous day in preparation for their maneuvers.

On the same day, some 3,700 miles away, off the western Pacific island of Rota in the Northern Marianas, approximately five hundred to seven hundred melon-headed whales were seen to congregate in Sasanhaya Bay, but they did not appear particularly agitated, and there were no strandings (Jefferson et al. 2006). Because there was a full moon that night, some observers tried to link the Kauai event with the Rota event on a lunar basis, but because the whales were probably behaving normally, that is, coming into the shallows for a rest, and there was no military activity anywhere near Rota, no connection can be drawn between the two stranding events. The likelihood remains that "the primary reason the whales entered and remained in Hanalei Bay until humans induced them to leave was the use of Mid-Frequency Sonar on 3 July 2004" (Brownell, Ralls, Baumann-Pickering, and Poole, 2009).

At Loza Lagoon, northwest Madagascar, in May to June 2008, one hundred melon-headed whales mass-stranded in a shallow tidal estuarine system, a most unusual location for this typically open-ocean dolphin. At first the cause of the mass stranding was unknown, but in 2013, an independent review panel including experts from the Wildlife Conservation Society, the International Fund for Animal Welfare, NOAA, the International Whaling Commission, and the government of Madagascar confirmed that the mass stranding was likely the result of a multibeam echo sounder system operated by a survey vessel contracted by ExxonMobil Exploration and Production (Northern Madagascar) Limited. The report (Southall et al. 2013) concludes:

> This is the first known such marine mammal mass stranding event closely associated with relatively high-frequency sonar systems. However, this alone is not a compelling reason to exclude the potential that the MBES [multibeam echosounder system] played a role in this event. Earlier such events may have been undetected because detailed inquiries were not conducted, given assumptions that high-frequency systems were unlikely to have such effects because of relatively greater sound propagation loss at high frequencies . . . Similar MBES systems to the 12 kHz source used in this case are in fact commonly used in hydrographic surveys around the world over large areas without such events being previously documented. In fact, a very similar MBES system was apparently used in a survey in the general area (and particularly the Mahajanga harbor area to the south) for some period during April and early-mid May 2008. This in fact could have played some contributing factor by sensitizing animals in the vicinity to such sources, but information on where and how this system was used was unavailable despite efforts to obtain it.

Long before sonar, loud sounds were used by hunters to drive small cetaceans into the shallows or even onto the shore, where they could be killed. Indeed, many "fisheries" depended on this technique, which can be dated as far back as the tenth-century Icelandic harvest of pilot whales and later, the pilot whale fisheries in Cape Cod, Newfoundland, Ireland, and the Faroe Islands. (The *grind* fishery in the Faroes is still going on.) In the Solomon Islands, dolphins were driven to the shallows by men clanging stones underwater. They killed the dolphins for their teeth, to be used in bride-price necklaces (Dawbin 1966). Japan has a centuries-old tradition of "drive-fishing," the most notorious example of which is the mass slaughter of bottlenose dolphins at Taiji, Japan, where despite the terrible publicity engendered by the Academy Award–winning film *The Cove*, the fishery is still being carried out annually. If you wanted to drive cetaceans into the shallows or to the shore, the best way to do it would be to make a lot of noise where they had no place to go but onto the shore, which is exactly what happened to the dolphins at Falmouth Bay. Common dolphins are not beaked whales, of course, but like all toothed cetaceans, their sense of hearing is particularly sensitive, and loud noises are likely to affect their behavior—and not for the better.

The British developed their own version of sonar during World War II, also to track enemy submarines. They called it ASDIC (Anti-Submarine Detection Investigation Committee) after the engineers that had developed it, and after the war, the technology was appropriated by the whaling industry to improve the efficiency of the hunt. Soon British and Japanese whaleships were using ASDIC to keep track of the whales and also to frighten them. A Norwegian whaler named Arne Skontorp developed a device he called the "whale scarer," which broadcast high-frequency pulses designed to drive whales to the surface and "induce panic and panting to fatigue the whale as quickly as possible" (Brownell, Nowacek, and Ralls 2008). The Japanese, who had launched themselves into the forefront of modern whaling technology by the 1960s, found that catcher boats with ASDIC were more efficient than boats without, and "tens of thousands of large whales were killed with greater efficiency with ASDIC. . . . Japanese sperm whale operations using ASDIC killed over fifty thousand animals in the North Pacific (Ohsumi 1980), and over seventeen thousand sperm whales were landed in Durban, South Africa [by the South Africa whaling fleet], after 1966, when almost the entire fleet was fitted with ASDIC" (Brownell et al. 2008). Purposely directed at hunted whales, ASDIC (or sonar) had a deleterious effect on the whales, as the sounds made it easier for the whalers to find and kill them. Accidentally deployed, say, in naval exercises where the sounds are not aimed at the whales, sonar can have an even more damaging effect, as the actual sounds can disorient, injure, or even kill the whales.

Before sonar, the only known threats to the deep-diving beaked whales were killer whales, cookiecutter sharks, and in some areas, small-boat whalers who hunted them for their oil. A new enemy has now appeared on the scene, more formidable than the little sharks, and far more dangerous than Norwegian or Japanese harpooners: the U.S. Navy. Of course, it is not the Navy's policy to kill marine mammals, but the repeated deployment of tactical sonar systems in regions occupied by beaked whales is tantamount to an attack on the whales. In virtually every instance where a number of whales have been found stranded, the Navy has first denied that they were there, citing the need for military secrecy, then claimed that sonar had nothing to do with the death of the whales. It can hardly be a coincidence that every time a number of beaked whales has fetched up dead on an island beach somewhere, naval maneuvers were being conducted nearby.

DeRuiter et al. (2013) observed that Cuvier's beaked whales stopped feeding, stopped swimming, and then swam rapidly away from the noise, often making unusually deep—but nonfeeding—dives. In addition to sonar, noise pollution by induced seismology blasts, used to locate buried oil and gas deposits, and the steady increase in commercial ship traffic, adds to the background noise intensity and adversely affects the welfare of cetaceans that depend on sound to survive. We know that mid-fequency military sonar affects beaked whales, but until researchers were able to test the responses of baleen whales, no one knew how they would react to the noise. Off Southern California, when deep-diving blue whales were subjected to simulated mid-frequency sonar, even below the range of operational military systems, they ceased deep feeding for hours at a time, increased their swimming speed, and swam away from the sound source. As Goldbogen et al. wrote in a 2013 study, "Sound-induced disruption of feeding and displacement from high-quality prey patches could have significant and previously undocumented impacts on baleen whale foraging ecology, individiual fitness and population health."

On January 16, 2008, President George W. Bush exempted the U.S. Navy from the law and argued that naval exercises are crucial to national security. A federal judge overruled the presidential exemption, but then a three-judge federal appeals court ruled that the Navy only had to take "precautions" during their sonar operations. After the fact, the Spanish Ministry of Defense announced a prohibition on all active sonar activities off the Canary Islands, and the U.S. ceased sonar training in the Bahamas, but the U.S. Navy opposes "any international regulatory framework addressing military use of active sonar, no matter what the science may now or in the future suggest" (Jansy 2005). According to an official U.S. Navy website: "Historical records show that marine mammal strandings have taken place for centuries, well before the advent of sonar. Nonetheless, the Navy remains dedicated to improving the collective understanding of the effects of sonar on marine mammals and to minimizing adverse effects of sonar consistent with our responsibility to defend the nation and ensure the safety of our Sailors, Airmen and Marines." In other words, it's better to be safe than sorry for the whales.

So in late 2012, the Navy issued a "Draft Environmental Impact Statement" with regard to the Hawaii–Southern California Testing and Training

Range, which incorporates vast swaths of the eastern Pacific, from northern Washington to Baja California, and includes offshore islands such as the Channel Islands off Southern California and the waters of the Hawaiian Islands. Based on the Navy's own analysis, testing in these areas will lead to millions of cases of hearing loss in marine mammals, more than five thousand serious injuries, and at least one thousand deaths. The National Marine Fisheries Service is prepared to give the Navy a permit to use high-powered explosives and deafening sonar during five years of training and testing—with devastating impacts on whales, dolphins, and other marine mammals.

On July 31, 2013, the U.S. Navy wrote to the California Coastal Commission, officially rejecting suggestions that sonar training be limited to avoid harming whales and other marine mammals in California waters. The commission had asked the Navy to stop training at times when certain whales are known to be present; blue whales arrive in California waters in August and September. (Various dolphins and beaked whales are in these waters year-round but are apparently not as charismatic as blue whales.) The Navy plans to continue sonar and underwater explosives training in Southern California and Hawaii between 2014 and 2019 and contends that the protections now in place are adequate.

Organized by Britain's Joint Tactical Exercise Planning Staff, "Exercise Joint Warrior" is a military exercise held twice a year that includes about thirteen thousand military personnel, forty military aircraft, and forty-nine ships from the United Kingdom, Canada, the Netherlands, Belgium, Sweden, Germany, Brazil, and the United States. It took place across Scotland from April 15 to 29, 2013, and includes airborne assault and amphibious landings as well as training in counter-insurgency, counter-piracy, and interstate war. Live ammunition is used in certain exercises, on cleared ranges where it poses no risk. One of the premier components of naval warfare, real or staged, is sonar.

The differences between beaked whales and other whales (and all dolphins, no matter the size) are substantial enough to place them in a category of their own, the Ziphiidae, but no matter what taxonomists call them, they are certainly whales. Because of their specialized anatomy and distincive habits, the beaked whales are patricularly vulnerable to being harmed by military sonar, but this represents only the latest chapter in our war on whales. For a thousand years, men killed whales for their oil, baleen, meat, and, often as a side industry, their ivory teeth. Although a few retrograde nations persist in the anachronistic business of commercial whale killing, for the most past, whaling for profit has come to a merciful end. And yet . . . in a 2013 issue of the journal *Conservation Biology*, van der Hoop and colleagues published a detailed report they called "Assessment of Management to Mitigate Anthropogenic Effects on Large Whales" (translation: people are responsible for the deaths of large whales), in which they looked at 1,763 known deaths and fatal injuries in the northwest Atlantic from 1970 to 2009 and concluded that 502 (67%) resulted from human interactions. Those that could not be attributed to natural causes were caused by entanglement in fishing gear and ship strikes. The authors concluded, "So far, regulatory efforts have not reduced the lethal effects of human activities to large whales on a population-range basis." So intentional or accidental, military or civilian, people are still killing whales, large and small, beaked and beakless.

Bibliography

Adams, R. D. 1979. *T*-phase recordings at Raratonga from undergound nuclear explosions. *Geophysical Journal of the Royal Astronomical Society* 58: 361–369.

Aguayo-L. A., D. Aurioles, J. Urbán-R., M. Salinez-Z., O. Vidal, and L. T. Findley. 1988. Beaked whales in Mexican waters. *Scientific Reports of the International Whaling Commission* SC/40/SM13: 1–9.

Aguilar de Soto, N., M. Johnson, P. T. Masden, P. L. Tyack, A. Bocconcelli, and A. Borsani. 2006. Does intense ship noise disrupt foraging in deep-diving Cuvier's beaked whales (*Ziphius cavirostris*)? *Marine Mammal Science* 22(3): 690–699.

Aguilar de Soto N., P. T. Madsen, P. Tyack, P. Arranz, J. Marrero, A. Fais, E. Revelli, and M. P. Johnson. 2011. No shallow talk: Cryptic strategy in the vocal communication of Blainville's beaked whales. *Marine Mammal Science* 28(2): E75–E92.

Aitken, P. F. 1971. Whales from the coast of South Australia. *Transactions of the Royal Society of South Australia* 95(2): 95–105.

Allen, A. N., J. J. Schanze, A. R. Solow, and P. L. Tyack. 2013. Analysis of a Blainville's beaked whale's movement response to playback of killer whale vocalizations. *Marine Mammal Science*. doi: 10.111/mms./12028.

Allen, B. M., J. G. Mead, and R. L. Brownell. 2011. Species review of Blainville's beaked whale. *Mesoplodon densirostris*. Paper SC/63/SM16 presented to the IWC Scientific Committee, May 2011 (unpublished). 18 pp.

Allen, G. M. 1939. True's beaked whale in Nova Scotia. *Journal of Mammalogy* 20(2): 259–260.

Anderson, R. C., R. Clark, P. T. Madsen, C. Johnson, J. Kiszka, and O. Breysse. 2006. Observations of Longman's beaked whale (*Indopacetus pacificus*) in the Western Indian Ocean. *Aquatic Mammals* 32(2): 223–231.

Andrews, R. C. 1908. Description of a new species of *Mesoplodon* from Canterbury Province, New Zealand. *Bulletin of the American Museum of Natural History* 24: 203–215.

———. 1911. Shore-whaling: A world industry. *National Geographic* 22(5): 411–442.

———. 1914. Note of a rare ziphoid whale, *Mesoplodon densirostris* on the New Jersey Coast. *Proceedings of the Academy of Natural Sciences of Philadelphia* 1914: 437–440.

———. 1931. *Whale hunting with gun and camera—a naturalist's account of the modern shore-whaling industry, of whales and their habits, and of hunting experiences in various parts of the world*. D. Appleton and Company, New York.

Anon. 1978. Bering Sea beaked whale. In "Alaska Whales and Whaling." *Alaska Geographic* 5(4): 72–73.

Aristotle. 1979. *Historia Animalium*. Loeb Classical Library. Harvard University Press.

Arnaya, I. N., N. Sano, and K. Iida. 1988. Studies on acoustic target strength of squid. I. Intensity and energy target strengths. *Bulletin of the Faculty of Fisheries Hokkaido University* 39(3): 187–200.

Arranz, P., N. A. de Soto, P. T. Madsen, A. Brito, F. Bordes, and M. P. Johnson. 2011. Following a foraging fish-finder: Diel habitat use of Blainville's beaked whales revealed by echolocation. *PLoS One* 6(12): e2853. doi.10.1371/ journal.pone.0028353.

Auster, P. J., and L. Watling. 2010. Beaked whale foraging areas inferred by gouges in the seafloor. *Marine Mammal Science* 26(1): 226–233.

Azzaroli, M. L. 1968. Second specimen of *Mesoplodon pacificus*, the rarest living beaked whale. *Monitore Zoologico Italiano* 2: 67–79.

Backus, R. H., and W. E. Schevill. 1961. The stranding of a Cuvier's beaked whale (*Ziphius cavirostris*) in Rhode Island, U.S.A. *Norsk Hvalfangst-tidende* 50(5): 177–181.

Baird, R. W. 2009. Beaked whale science coming of age. *Whalewatcher* 38(1): 18–22.

Baird, R. W., D. L. Webster, J. M. Aschettino, G. S. Schorr, and D. J. McSweeney. 2013. Odontocete cetaceans around the main Hawaiian Islands: Habitat use and relative abundance from small-boat sighting surveys. *Aquatic Mammals* 39: 253–269.

Baird, R. W, D. L. Webster, D. J. McSweeney, A. D. Ligon, G. S. Schorr, and J. Barlow. 2006. Diving behaviour of Cuvier's (*Ziphius cavirostris*) and Blainville's (*Mesoplodon densirostris*) beaked whales in Hawai'i. *Canadian Journal of Zoology* 84(8): 1120–1128.

Baird, R. W, D. L. Webster, G. S. Schorr, D. J. McSweeney, and J. Barlow. 2008. Diel variation in beaked whale diving behavior. *Marine Mammal Science* 24: 630–642.

Baker, A. N. 1972. New Zealand whales and dolphins. *Tuatara* 20(1): 1–49.

———. 1983. *Whales and Dolphins of New Zealand and Australia: An Identification Guide*. Victoria University Press.

———. 2001. Status, relationships, and distribution of *Mesoplodon bowdoini* Andrews, 1904. *Marine Mammal Science* 17(3): 473–493.

Baker, A. N., P. Duignan, R. J. de B. Norman and A. Van Helden. 2001. A juvenile Hector's beaked whale, *Mesoplodon hectori* (Gray 1871) without throat grooves, plus notes on endoparasites (Cetacea: Ziphiidae). *Marine Mammal Science* 17(1): 171–175.

Baker, A. N., and A. L. van Helden. 1999. New records of beaked whale, genus *Mesoplodon*, from New Zealand (Cetacea: Ziphiidae). *Journal of the Royal Society of New Zealand* 29(3): 235–244.

Balcomb, K. C. 1989. Baird's Beaked Whale—*Berardius bairdii* Stejneger, 1983; Arnoux's Beaked Whale—*Berardius arnuxii* Duvernoy, 1851. In S. H. Ridgway and R. Harrison, eds., *Handbook of Marine Mammals. Vol. 4: River Dolphins and the Larger Toothed Whales*, pp. 261–288. Academic Press.

Balcomb, K. C., and D. E. Claridge. 2001. A mass stranding of cetaceans caused by naval sonar in the Bahamas. *Bahamas Journal of Science* 8(2): 2–12.

Balcomb, K. C., and C. A. Goebel. 1977. Some information on a *Berardius bairdii* fishery in Japan. *Report of the International Whaling Commission* 27: 485–486.

Ballance, L. T., R. C. Anderson, R. L. Pitman, K. Stafford, A. Shaan, Z. Waheed, and R. L. Brownell. 2001. Cetacean sightings around the Republic of the Maldives, April 1998. *Journal of Cetacean Research and Management* 3(2): 213–218.

Ballance, L. T., and R. L. Pitman. 1998. Cetaceans of the western tropical Indian Ocean: Distribution, relative abundance, and comparisons with cetacean communities of two other tropical ecosystems. *Marine Mammal Science* 14(3): 429–459.

Barham, E. G. 1966. Deep-scattering layer migration and composition: Observations from a diving saucer. *Science* 151(3716): 1399–1403.

Barlow, J., and G. A. Cameron. 2003. Field experiments show that acoustic pingers reduce marine mammal bycatch in the California drift gillnet fishery. *Marine Mammal Science* 19(2): 265–283.

Barlow, J., and R. Gisiner. 2006. Mitigating, monitoring and assessing the effects of anthropogenic sound on beaked whales. *Journal of Cetacean Research and Management* 7(3): 239–249.

Barlow, J., M. C. Ferguson, W. F. Perrin, L. Ballance, T. Gerodette, G. Joyce, C. D. MacLeod, K. Mullin, D. L. Palka, and G. Waring. 2006. Abundance and densities of beaked and bottlenose whales (family Ziphiidae). *Journal of Cetacean Research and Management* 7(3): 263–270.

Barnett, A. 1997. Pouting whales are suckers for squid. *New Scientist* 2067: 17.

Bauman-Pickering, S., M. A. McDonald, A. E. Simonis, A. S. Berga, K. P. B. Merkens, E. M. Oleson, M. A. Roch, S. M. Wiggins, S. Rankin, T. M. Yack, and J. A. Hildebrand. 2013. Species-specific beaked whale echolocation signals. *Journal of the Acoustical Society of America* 134(3): 2293–2301.

Baumann-Pickering, S., M. A. Roch, R. L. Brownell Jr., A. E. Simonis, M. A. McDonald, A. Solsona-Berga, E. M. Oleson, S. M. Wiggins, and J. A. Hildebrand. 2014. Spatio-temporal patterns of beaked whale echolocation signals in the North Pacific. *PLoS ONE.* http:\\dx.doi.org\10.1371\journal.pone.0086072.

Baumann-Pickering, S., S. M. Wiggins, E. Roth, M. A. Roch, H.-U. Schnitzler, and J. A. Hildebrand. 2010. Echolocation signals of a beaked whale at Palmyra Atoll. *Journal of the Acoustical Society of America* 127: 3790–3799.

Baumann-Pickering, S., T. M. Yack, J. Barlow, S. M. Wiggins, and J. A. Hildebrand. 2013. Baird's beaked whale echolocation signals. *Journal of the Acoustical Society of America* 133(6): 4321–4331.

Beale, T. 1835. *A Few Observations on the Natural History of the Sperm Whale.* Effingham Wilson, London.

Beddard, F. E. 1900. *A Book of Whales.* John Murray.

Bel'kovich, V. M., and A. V. Yablokov. 1963. The whale—an ultrasonic projector. *Yuchnyi Teknik* 3: 76–77.

Beneden, P. J. v., and P. Gervais. 1868–1879. *Ostéographie des cétacés, vivants et fossiles, comprenant la description et l'iconographie du squelette et du système dentaire de ces animaux, ainsi que des documents relatifs a leur histoire naturelle.* Paris, A. Bertrand.

Benjaminsen, T. 1972. On the biology of the bottlenose whale *Hyperoodon ampullatus* (Foster). *Norwegian Journal of Zoology* 20: 233–241.

Benjaminsen, T., and I. Christensen. 1979. The natural history of the bottlenose whale, *Hyperoodon ampullatus* (Forster). In H. E. Winn and B. L Olla, eds., *Behavior of Marine Mammals. Vol. 3: Cetaceans*, pp. 143–164. Plenum Press.

Benoit-Bird, K. J., W. W. L. Au, and R. Kastelein. 2006. Testing the odontocete acoustic prey debilitation hypothesis: No stunning results. *Journal of the Acoustical Society of America* 120(2): 1118–1123.

Benoit-Bird, K. J., W. F. Gilly, W. W. L. Au, and B. Mate. 2008. Controlled and *in situ* target strengths of the jumbo squid *Dosidicus gigas* and the identification of potential acoustic scattering sources. *Journal of the Acoustical Society of America* 123(3): 1318–1328.

Berta, A. 1994. What is a whale? *Science* 263: 180–81.

Berta, A., and J. L. Sumich. 1999. *Marine Mammals: Evolutionary Biology*. Academic Press.

Berzin, A. A. 1972. *The Sperm Whale*. Izdatgel'stvo "Pischevaya Promyshlennost" Moskva 1971. Translated from the Russian by Israel Program for Scientific Translation, Jerusalem.

Berzin, A. A., and G. M. Veinger. 1976. Investigations of the population morphology of sperm whales, *Physeter macrocephalus* L. 1758 of the Pacific Ocean. *FAO Symposium on Marine Mammals*. ACMRR/MM/SC135.

Besharse, C. 1971. Maturity and sexual dimorphism in the skull, mandible and teeth of the beaked whale, *Mesoplodon densirostris*. *Journal of Mammalogy* 52(2): 297–315.

Best, P. B., I. J. Ansorge, J. R. E. Lutjeharms, W. F. Perrin, and P. A. Folkens (ill.). 2007. *Whales and Dolphins of the Southern African Subregion*. Cambridge University Press, Cape Town.

Bianucci, G., W. Landini, and A. Varola. 1992. *Messapicetus longirostris*, a new genus and species of Ziphiidae (Cetacea) from the late Miocene of "Pietra leccese" (Apulia, Italy). *Bollettino della Societa Paleontologica Italiana* 31(2): 261–264.

Bianucci G., I. Miján, O. Lambert, K. Post, and O. Mateus, 2013. Bizarre fossil beaked whales (Odontoceti, Ziphiidae) fished from the Atlantic Ocean floor off the Iberian Peninsula. *Geodiversitas* 35(1): 105–153.

Bianucci, G., K. Post, and O. Lambert. 2008. Beaked whale mysteries revealed by seafloor fossils trawled off South Africa. *South African Journal of Science* 104: 140–142.

Blainville, H. M. D. de. 1817. In "Dauphins," *Nouveau Dictionaire d'Histoire Naturelle, appliquée aux arts, à l'agriculture, à l'économie rurale et domestique, à la médecine, etc.* 9: 178–179.

Blanco, C., and J. A. Raga. 2000. Cephalopod prey of two *Ziphius cavirostris* (Cetacea) stranded on the western Mediterranean coast. *Journal of the Marine Biological Association of the U.K.* 80: 381–382.

Bloodworth, B., and C. D. Marshall. 2005. Feeding kinematics of *Kogia* and *Tursiops* (Odontoceti: Cetacea): Characterization of suction and ram feeding. *Journal of Experimental Biology* 208: 3721–3730.

Bonde, R. K., and T. J. O'Shea. 1989. Sowerby's beaked whale (*Mesoplodon bidens*) in the Gulf of Mexico. *Journal of Mammalogy* 70(2): 447–448.

Boschma, H. 1950. Maxillary teeth in specimens of *Hyperoodon rostratus* (Muller) and *Mesoplodon grayi* von Haast stranded on the Dutch coasts. *Koninklike Nederlandse Akademie van Wetenschapper, Proceedings* 53(6): 775–786.

———. 1951a. Rows of small teeth in ziphoid whales. *Zoologische Mededelingen Leiden* 31: 139–148.

———. 1951b. Some smaller whales. *Endeavor* 10(39): 131–135.

Brazenor, C. W. 1933. First record of a beaked whale (*Mesoplodon grayi*) from Victoria. *Proceedings of the Royal Society of Victoria* 45(1): 23–24.

Brill, R. L., M. L. Sevenich, T. J. Sullivan, J. D. Sustman, and R. E. Witt. 1988. Behavioral evidence for hearing through the lower jaw by an echolocating dolphin (*Tursiops truncatus*). *Marine Mammal Science* 4: 223–230.

Brimley, H. H. 1943. A second specimen of True's beaked whale *Mesoplodon mirus* from North Carolina. *Journal of Mammalogy* 24(2): 199–203.
Brownell, R. L. 1974. Small odontocetes of the Antarctic. In V. C. Bushnell, ed., *Antarctic Mammals*, folio 18, pp. 13–19. Antarctic Map Folio Series, American Geographic Society.
Brownell, R. L., A. Aguayo, and D. N. Torres. 1976. A Shepherd's beaked whale, *Tasmacetus shepherdi*, from the eastern South Pacific. *Scientific Reports of the Whales Research Institute, Tokyo* 28: 127–128.
Brownell, R. L., D. P. Nowacek, and K. Ralls. 2008. Hunting cetaceans with sound: A worldwide review. *Journal of Cetacean Research and Management* 10: 81–88.
Brownell, R. L., K. Ralls, S. Baumann-Pickering, and M. M. Poole. 2009. Behavior of melon-headed whales, *Peponocephala electra*, near oceanic islands. *Marine Mammal Science* 25(3): 639–658.
Bruemmer, F. 1993. *The Narwhal: Unicorn of the Sea.* Canadian Geographic/Key Porter Books.
Budelmann, B-.U. 1980. Equilibrium and orientation in cephalopods. *Oceanus* 23(3): 34–43.
———. 1990. The statocysts of squid. In D. L. Gilbert, W. J. Adelman, and J. M. Arnold, eds., *Squid as Experimental Animals*, pp. 421–442. Plenum.
———. 1992. Hearing in nonarthropod invertebrates. In D. B. Webster, R. R. Fay, and A. N. Popper, eds., *The Evolutionary Biology of Hearing*, pp. 141–55. Springer-Verlag.
Budelmann, B-.U., and H. Bleckmann. 1988. A lateral line analogue in cephalopods: Water waves generate microphonic potentials in the epidermal head lines of *Sepia* and *Lollinguncula*. *Journal of Comparative Physiology* 164: 1–5.
Buono, M., and M. A. Cozzuol. 2013. A new beaked whale (Cetacea, Odontoceti) from the Late Miocene of Patagonia, Argentina. *Journal of Vertebrate Paleontology* 33(4): 986–997.
Caldwell, D. K., and M. C. Caldwell. 1971. Sounds produced by two rare cetaceans stranded in Florida. *Cetology* 4: 1–6.
Cappozzo, H. L., M. F. Negri, B. Mahler, V. V. Lia, P. Martinez, A. Gianggiobe, and A. Saubidet. 2005. Biological data on two Hector's beaked whales, *Mesoplodon hectori*, stranded in Buenos Aires province, Argentina. *Latin American Journal of Aquatic Mammals* 4: 113–128.
Caretta, J., J. Barlow, and L. Enriquez. 2008. Acoustic pingers eliminate beaked whale bycatch in gill net fishery. *Marine Mammal Science* 24(4): 956–961.
Carlström, J., J. Denkinger, P. Feddersen, and N. Øien. 1997. Record of a new northern range of Sowerby's beaked whale (*Mesoplodon bidens*). *Polar Biology* 17(5): 459–461.
Christensen, I. 1973. Age determination, age distribution and growth of bottlenose whale *Hyperoodon ampullatus* in the Labrador Sea. *Norwegian Journal of Zoology* 21: 331–340.
———. 1975. Preliminary report on the Norwegian fishery for small whales: Expansion of Norwegian whaling to Arctic and northwest Atlantic waters, and Norwegian investigations of the biology of small whales. *Journal of the Fisheries Research Board of Canada* 32(7): 1083–1094.
Clark, E. 1969. *The Lady and the Sharks.* Harper and Row.
Clarke, M. R. 1962. Stomach contents of a sperm whale caught off Madeira in 1959. *Norsk Hvalfangst-Tidende* 5: 173–191.

———. 1970. Function of the spermaceti organ of the sperm whale. *Nature* 228: 873–874.
———. 1976. Observations on sperm whale diving. *Journal of the Marine Biological Association of the U.K.* 56: 809–810.
———. 1978a. Structure and proportions of the spermaceti organ in the sperm whale. *Journal of the Marine Biological Association of the U.K.* 58: 1–17.
———. 1978b. Physical properties of spermaceti oil in the sperm whale. *Journal of the Marine Biological Association of the U.K.* 58: 19–26.
———. 1978c. Buoyancy control as a function of the spermaceti organ in the sperm whale. *Journal of the Marine Biological Association of the U.K.* 58: 27–71.
———. 1979. The head of the sperm whale. *Scientific American* 240(1): 128–141.
Clarke, R. 2005. A southern bottlenose whale examined in the Antarctic. *Latin American Journal of Aquatic Mammals* 4(2): 83–96.
Colbert, E. H. 1955. *The Evolution of the Vertebrates.* John Wiley and Sons.
Cowan, I. M. 1945. A beaked whale stranded on the coast of British Columbia. *Journal of Mammalogy* 26(1): 93–94.
Cox, T. M., T. J. Ragen, A. J. Read, E. Vos, R. W. Baird, K. Balcomb, J. Barlow, J. Caldwell, T. Cranford, L. Crum, A. D'Amico, G. D'Spain, A. Fernandez, J. Finneran, R. Gentry, W. Gerth, F. Gulland, J. Hildebrand, D. Houser, P. D. Jepson, D. Ketten, C. D. MacLeod, P. Miller, S. Moore, D. C. Mountain, D. Palka, P. Poganis, S. Rommel, T. Rowles, B. Taylor, P. Tyack, D. Wartzok, R. Gisiner, J. Mead, and L. Benner. 2006. Understanding the impacts of anthropogenic sound on beaked whales. *Journal of Cetacean Research and Management* 7(3): 177–187.
Cozzi, B., M. Panin, C. Butti, M. Podesta, and A. Zotti. 2010. Bone density distribution patterns in the rostrum of delphinids and beaked whales: Evidence of family specific evolutive traits. *Anatomical Record* 293: 235–242.
Cranford, T. W., M. F. McKenna, M. S. Soldevilla, S. M. Wiggins, J. A. Goldbogen, R. E. Shadwick, P. Krysl, J. A. St. Leger, and J. A. Hildebrand. 2008. Anatomic geometry of sound transmission and reception in Cuvier's beaked whale (*Ziphius cavirostris*). *Anatomical Record* 291: 353–378.
Cummings, W. C., J. F. Fish, and P. O. Thompson. 1972. Sound production and other behavior of southern right whales, *Eubalaena glacialis. Transactions of the San Diego Society of Natural History* 17(1): 1–13.
Cuvier, G. 1824. *Recherches sur les Ossements Fossiles.* Éd. 4, vol. 8, pp. 153–321. Paris.
Dakin, W. J. 1938. *Whalemen Adventurers.* Angus and Robertson.
Dalebout, M. L., C. S. Baker,. J. G. Mead, V. G. Cockcroft, and T. K. Yamada. 2004. A comprehensive and validated molecular taxonomy of beaked whales, family Ziphiidae. *Journal of Heredity* 95(6): 459–473.
Dalebout, M. L., C. S. Baker, D. Steel, K. M. Robertson, S. J. Chivers, W. F. Perrin, J. G. Mead, R. V. Grace, and T. D. Schofield. 2007. A divergent mtDNA lineage among *Mesoplodon* beaked whales: Molecular evidence for a new species in the tropical Pacific? *Marine Mammal Science* 23(4): 954–966.
Dalebout, M. L., C. S. Baker, D. Steel, K. Thompson, K. M. Robertson, S. J. Chivers, W. F. Perrin, M. Goonatilake, R. C. Anderson, J. G. Mead, C. W. Potter, T. K. Yamada, L. Thompson, and D. Jupiter. 2012. A newly recognised beaked whale (Ziphiidae) in the Tropical Indo-Pacific: *Mesoplodon hotaula* or *M. ginkgodens hotaula.* IWC SC/64/SM3.

Dalebout, M. L., J. G. Mead, C. S. Baker, A. N. Baker, and A. L. Van Helden. 2002. A new species of beaked whale *Mesoplodon perrini* sp. n. (Cetacea: Ziphiidae) discovered through phylogenetic analyses of mitochondrial DNA sequences. *Marine Mammal Science* 18: 577–608.

Dalebout, M. L., G. J. B. Ross, C. S. Baker, R. C. Anderson, P. B. Best, V. G. Cockcroft, H. L. Hinsz, V. Peddemors, and R. L. Pitman. 2003. Appearance, distribution, and genetic distinctiveness of Longman's beaked whale, *Indopacetus pacificus. Marine Mammal Science* 19(3): 421–461.

Dalebout, M. L., K. G. Russell, M. J. Little, and P. Ensor. 2004. Observations of live Gray's beaked whales (*Mesoplodon grayi*) in Mahurangi Harbor, North Island, New Zealand, with a summary of at-sea sightings. *Journal of the Royal Society of New Zealand* 34(4): 347–356.

Dalebout, M., D. Steele, and C. S. Baker. 2008. Phylogeny of the beaked whale genus *Mesoplodon* (Ziphiidae: Cetacea) revealed by nuclear introns: Implications for the development of male tusks. *Systematic Biology* 57(6): 857–875.

Dalebout, M. L., A. L. Van Helden, K. Van Waerbeek, and C. S. Baker. 1998. Molecular genetic identification of southern hemisphere beaked whales (Cetacea: Ziphiidae). *Molecular Ecology* 7: 687–694.

D'Amico, A., R. C. Gisiner, D. R. Ketten, J. A. Hammock, C. Johnson, P. L. Tyack, and J. G. Mead (2009). Beaked whale strandings and naval exercises. *Aquatic Mammals* 55(4): 452–472.

Darwin, C. 1859. *On the Origin of Species*. John Murray, London.

Davis, R. W., N. Jaquet, D. Gendron, U. Markaida, G. Bazzino, and W. Gilly. 2007. Diving behavior of sperm whales in relation to behavior of a major prey species, the jumbo squid, in the Gulf of California, Mexico. *Marine Ecology Progress Series* 333: 291–302.

Dawbin, W. H. 1966. Porpoises and porpoise hunting in Malaita. *Australian Natural History* 15(7): 207–211.

Dawson, S. M., J. Barlow, and D. Ljungblad. 1998. Sounds recorded from Baird's beaked whale, *Berardius bairdii. Marine Mammal Science* 14(2): 335–344.

De Buffrénil, V., and O. Lambert. 2011. Histology and growth pattern of the pachyosteosclerotic premaxillae of the fossil beaked whale *Aporotus recurvirostris* (Mammalia, Cetacea, Odontoceti). *Geobios* 44(1): 45–56.

De Buffrénil, V., L. Zylberberg, W. Traub, and A. Casinos. 2000. Structural and mechanical characteristics of the hyperdense bone of the rostrum of *Mesoplodon densirostris* (Cetacea, Ziphiidae): Summary of recent observations. *Historical Biology* 14: 57–65.

Deecke, V. B., J. K. B. Ford, and P. J. B. Slater. 2005. The vocal behaviour of mammal-eating killer whales: communicating with costly calls. *Animal Behavior* 69: 395–405.

Deraniyagala, P. E. P. 1963a. A new beaked whale from Ceylon. *Ceylon Today* 12: 13–14.

———. 1963b. Mass mortality of the new subspecies of little piked whale *Balaenoptera acutorostrata thalmaha* and a new beaked whale *Mesoplodon hotaula* from Ceylon. *Spolia Zeylanica* 30(1): 79–84.

———. 1963c. Comparison of *Mesoplodon hotaula* Deraniyagala with *Ziphius cavirostris indicus* (van Beneden). *Spolia Zeylanica* 30(2): 248–256.

DeRuiter, S. D., B. L. Southall, J. Calambokidis, W. M. X. Zimmer, D. Sadykova, E. A. Falcone, A. R. Friedlander, J. E. Joseph, D. J. Moretti, G. S. Schorr, A. Douglas, L. Thomas, and P. L. Tyack. 2013. First direct measurements of behavioural responses by Cuvier's beaked whales to mid-frequency active sonar. *Biology Letters* 9. http://dx.doi.org/10.1098/rsbl.2013.0223.

Dewhurst, W. H. 1834. *The Natural History of the Order Cetacea and the Oceanic Inhabitants of the Arctic Regions*. London.

Dietz, R., A. D. Shapiro, M. Bahktiari, J. Orr, P. L. Tyack, P. Richard, I. G. Eskesen, and G. Marshall. 2007. Upside-down swimming behaviour of free-ranging narwhals. *BMC Ecology* 7(14). doi:10.1186/1472-6785-7-14.

Dolman, S. J., P. G. H. Evans, G. Notabartolo-diScarra, and H. Frisch. 2010. Active sonar, beaked whales and European regional policy. *Marine Pollution Bulletin*. doi:10.1016/j.marpolbul.2010.03.034.

Dolman, S. J., C. D. MacLeod, and P. G. H. Evans. 2006. Adaptations and evolution of structure for intraspecific combat in the family Ziphiidae. Poster NH2, 20th Annual Conference of the European Cetacean Society, Gdynia, Poland, April 1–6, 2006.

Dolman, S. J., E. Pinn, R. J. Reid, J. P. Barley, R. Deaville, P. D. Jepson, M. O'Connell, S. Berrow, R. S. Penrose, P. T. Stevick, S. Calderan, K. P. Robinson, R. L. Brownell Jr., and M. P. Simmonds. 2010. A note on the unprecedented strandings of 56 deep-diving whales along the UK and Irish coast. *Marine Biodiversity Records*. doi:10.1017/S175526720999114X.

D'Spain, G. L., A. D'Amico, and D. M. Fromm. 2006. Properties of the underwater sound fields during some well documented beaked whale stranding events. *Journal of Cetacean Research and Management* 7(3): 223–238.

Dudley, P. 1725. An essay upon the natural history of whales, with particular account of the ambergris found in the sperma ceti whale. *Philosophical Transactions of the Royal Society of London* 33(387): 256–59.

Dudok van Heel, W. H. 1974. Remarks on a live ziphiid baby (*Mesoplodon bidens*). *Aquatic Mammals* 2(2): 3–7.

Duguy, R. 1977. Notes on the small cetaceans off the coast of France. *Report of the International Whaling Commission* 27: 500–501.

Dunn, C., L. Hickmott, D. Talbot, I. Boyd, and L. Rendell. 2013. Mid-frequency broadband sounds of Blainville's beaked whales. *Bioacoustics* 22(2): 153–163.

Duvernoy, [G. L.] (1851). "Memoire sur les charactères ostéologues des genres nouveaux ou des espèces nouvelles de cétacés vivants ou fossiles." *Annales des Sciences Naturelles, Troisième série, zoologie* 15: 5–71, pls. 71–72, figs. 71–75.

Ellis, R. 1980a. Beaked whales. *Sea Frontiers* 26(1): 10–18.

———. 1980b. *The Book of Whales*. Knopf.

———. 1981. Observations on a captive sperm whale, *Physeter macrocephalus*, at Fire Island, New York (abstract). *Fourth Biennial Conf. Biol. Marine Mammals*. San Francisco.

———. 1982. *Dolphins and Porpoises*. Knopf.

———. 1993. *Physty: The True Story of a Young Whale's Rescue*. Running Press.

———. 1998. *The Search for the Giant Squid*. Lyons.

———. 2001. *Aquagenesis: The Origin and Evolution of Life in the Sea*. Viking.

———. 2011. *The Great Sperm Whale*. University Press of Kansas.

Empeño, H. 2010. Elusive beaked whale stranded in Subic. *SubicNewsLink*.

Erdman, D. S. 1962. Stranding of a beaked whale *Ziphius cavirostris* Cuvier on the south coast of Puerto Rico. *Journal of Mammalogy* 43(2): 276–277.

Faerber, M. M., and R. W. Baird. 2010. Does a lack of observed beaked whale strandings in military exercise areas mean no impacts have occurred? A comparison of stranding and detection probabilities in the Canary and main Hawaiian Islands. *Marine Mammal Science* 26(3): 602–613.

Fahlke, J. M., P. D. Gingerich, R. C. Welsh, and A. R. Wood. 2011. Cranial asymmetry in Eocene archaeocete whales and the evolution of directional hearing in water. *Proceedings of the National Academy of Sciences* 108(35): 14545–14548.

Fernández A., M. Arbelo, R. Deaville, I. A. P. Patterson, P. Castro, J. R. Baker, E. Degollada, H. M. Ross, P. Herráez, A. M. Pocknell, E. Rodriguez, F. E. Howie, A. Espinosa, R. J. Reid, J. R. Jaber, V. Martín, A. A. Cunningham, and P. D. Jepson. 2004. Pathology: Whales, sonar, and decompression sickness. *Nature* 428: 1–2.

Fernández A., M. Arbelo, and V. Martín. 2013. No mass strandings since sonar ban. *Nature* 497: 317.

Flower, W. H. 1872. On the recent ziphoid whales, with a description of the skeleton of *Berardius arnouxi. Transactions of the Zoological Society of London* 8(3): 203–233.

———. 1882. On the whales of the genus *Hyperoodon. Proceedings of the Zoological Society of London* 1882: 722–726.

Ford, J. K. B., G. M. Ellis, and K. C. Balcomb. 1994. *Killer Whales.* University of British Columbia Press.

Ford, J. K. B., and R. R. Reeves. 2008. Fight or flight: Antipredator strategies of baleen whales. *Mammal Review* 38: 50–86.

Fordyce, R. E., and L. G. Barnes. 1994. The evolutionary history of whales and dolphins. *Annual Review of Earth and Planetary Sciences* 22: 419–55.

Frantzis, A. 1998. Does acoustic testing strand whales? *Nature* 392: 29.

———. 2004. The first mass stranding that was associated with the use of active sonar (Kyparissiakos Gulf, Greece, 1996). *Proceedings of the Workshop on Active Sonar and Cetaceans.* Las Palmas, Gran Canaria, March 8, 2003, pp. 14–20.

Fraser, F. C. 1934. Report on cetacea stranded on the British coasts from 1927 to 1932. *Bulletin of the British Museum (Natural History)* 11: 1–41.

———. 1945. On a specimen of the southern bottlenosed whale, *Hyperoodon planifrons. Discovery Reports* 23: 19–36.

———. 1946. Report on cetacea stranded on the British coasts from 1933 to 1937. *Bulletin of the British Museum (Natural History)*12: 1–56.

———. 1950. Notes on a skull of Hector's beaked whale *Mesoplodon hectori* (Gray) from the Falkland Islands. *Proceedings of the Linnaean Society of London* 162: 50–52.

———. 1953. Report on cetacea stranded on the British coasts from 1938 to 1947. *Bulletin of the British Museum (Natural History)* 13: 1–48.

———. 1955. A skull of *Mesoplodon gervaisi* (Deslongchamps) from Trinidad, West Indies. *Annals and Magazine of Natural History* 8(92): 624–630.

———. 1974. Report on cetacea stranded on the British coasts from 1948 to 1966. *Bulletin of the British Museum (Natural History)* 14: 1–65.

Fristrup, K. M., and G. R. Harbison. 2002. How do sperm whales catch squids? *Marine Mammal Science* 18(1): 42–54.

Fuller, A. J., and S. J. Godfrey. 2007. A late Miocene ziphiid (*Messapicetus* sp.: Odontoceti: Cetacea) from the St. Marys formation of Calvert Cliffs, Maryland." *Journal of Vertebrate Paleontology* 27(2): 535–540.

Galbreath, E. C. 1963. Three beaked whales stranded in the Midway Islands, Central Pacific Ocean. *Journal of Mammalogy* 44(3): 422–423.

Gales, N. J., M. L. Dalebout, and J. Bannister. 2002. Genetic identification and biological observation of two free-swimming beaked whales, Hector's beaked whale (*Mesoplodon hectori* (Gray, 1871) and Gray's beaked whale (*Mesoplodon grayi* Von Haast, 1876) *Marine Mammal Science* 18(3): 544–551.

Gallo-Reynoso, J., and A. Figuero-Carranza. 1995. Occurrence of bottlenose whales in the waters of Isla Guadalupe, Mexico. *Marine Mammal Science* 11(4): 573–575.

Gambell, R. 1970. Weight of a sperm whale, whole and in parts. *South African Journal of Science* 66: 225–227.

Gaskin, D. E. 1968. The New Zealand cetacea. *Fisheries Research Bulletin* 1: 1–92.

Gatesy, J., J. H. Geisler, J. Chang, C. Buell, A. Berta, R. W. Meredith, M. S. Springer, and M. R. McGowen. A phylogenetic blueprint for a modern whale. *Molecular Phylogenetics and Evolution* 66: 479–506.

Geist, V. 1966. The evolution of horn-like organs. *Behaviour* 27(3/4): 175–214.

———. 1998. *Deer of the World: Their Evolution, Behavior, and Ecology*. Stackpole.

Gero, S. 2012. The surprisingly familiar family lives of sperm whales. *Whalewatcher* 41(1): 16–20.

Gero, S., and H. Whitehead. 2007. Suckling behavior in sperm whale calves: Observations and hypotheses. *Marine Mammal Science* 23(2): 398–413.

Gianuca, N. M., and H. P. Castello. 1976. First record of the southern bottlenose whale, *Hyperoodon planifrons*, from Brazil. *Scientific Reports of the Whales Research Institute* 28: 119–126.

Gingerich, P. D. 1998. Paleobiological perspectives on Mesonychia, Archaeoceti, and the origin of whales. In J. G. M. Thewissen, ed., *The Emergence of Whales*, pp. 423–450. Plenum.

Glauert, L. 1947. The genus *Mesoplodon* in western Australian seas. *Australian Zoology* 11(2): 73–75.

Goldbogen, J. A., B. L. Southall, S. D. DeRuiter, J. Calambokidis, A. R. Friedlander, E. L. Hazen, E. A. Falcone, G. S. Schorr, A. Douglas, D. J. Moretti, C. Kyburg, M. F. McKenna, and P. L. Tyack. 2013. Blue whales respond to simulated mid-frequency military sonar. *Proceedings of the Royal Society B*. doi.org/10.1098/rspb.2013.0657.

Gol'din, P. E., and K. A. Vishnyakova. 2013. *Africanacetus* from the sub-Antarctic region: The southernmost record of fossil beaked whales. *Acta Paleontologica Polonica* 58(3): 445–452.

Goodall, R. N. P. 1978. Report on small cetaceans stranded on the coasts of Tierra del Fuego. *Scientific Reports of the Whales Research Institute* 30: 197–230.

———. 1988. The Hector's beaked whale, *Mesoplodon hectori*, off southern South America. *Scientific Reports of the International Whaling Commission* SC/40/SM18: 1–15.

Goodall, R. N. P., and A. N. Baker. 1988. Review of knowledge of the Shepherd's beaked whale, *Tasmacetus shepherdi*. *Scientific Reports of the International Whaling Commission* SC/40/SM18: 1–21.

Goodall, R. N. P., C. C. Boy, L. E. Pimper, and S. M. Nacnie. 2004. Range extensions and exceptional records of cetaceans for Tierra del Fuego. *Abstracts, 11° Reunión de Trabajo de Especialistas en Mamíferos Acuáticos de Américadel Sur y 5° Congreso de la SOLAMAC, 11–17 September 2004, Quito, Ecuador*, p. 158.

Goodall, R. N. P., C. L. Folger, and A. A. Lichter. 1988. The presence of the Layard's beaked whale, *Mesoplodon layardii*, in the southwest South Atlantic, with a review of strandings worldwide. *Scientific Reports of the International Whaling Commission* SC/40/SM20: 1–39.

Gottfried, M. D., D. J. Bohaska, and F. C Whitmore. 1994. Miocene cetaceans of the Chesapeake Group. In A. Berta and F. C. Whitmore, *Contributions in Marine Paleontology Honoring Frank C. Whitmore*, pp. 229–238. *Proceedings of the San Diego Society of Natural History* 29.

Gowans, S., M. Dalebout, S. K. Hooker, and H. Whitehead. 2000. Reliability of photographic and molecular techniques for sexing northern bottlenose whales (*Hyperoodon ampullatus*). *Canadian Journal of Zoology* 78: 1224–1229.

Gowans, S., and L. Rendell. 1999. Head-butting in northern bottlenose whales (*Hyperoodon ampullatus*): A possible function for big heads? *Marine Mammal Science* 15(4): 1342–1350.

Gowans, S., and H. Whitehead. 2001. Photographic identification of northern bottlenose whales (*Hyperoodon ampullatus*): Sources of heterogeneity from natural marks. *Marine Mammal Science* 17(1): 76–93.

Gowans, S., H. Whitehead, and S. K. Hooker. 2001. Social organization in northern bottlenose whales, *Hyperoodon ampullatus*: Not driven by deep-water foraging? *Animal Behaviour* 62(2): 369–377.

Grandi, M. F., A. D. Buren, E. A. Crespo, N. A. Garcia, G. M. Svendsin, and S. L. Dans. 2005. Record of a specimen of Shepherd's beaked whale (*Tasmacetus shepherdi*) from the coast of Santa Cruz, Argentina, with notes on age determination. *Latin American Journal of Aquatic Mammals* 4(2): 97–100.

Gray, D. 1882. Notes on the characters and habits of the bottle-nose whale (*Hyperoodon rostratus*). *Proceedings of the Zoological Society of London* 1882: 726–731.

Gray, J. E. 1860. On the genus *Hyperoodon*: The two British kinds and their food. *Proceedings of the Zoological Society of London* 28: 422–426.

———. 1865. Notes on the whales of the Cape; by E. L. Layard, Esq. Capetown Corr. Memb., with descriptions of two new species; by Dr. J. E. Gray. *Proceedings of the Zoological Society of London* 1865: 357–359.

———. 1871. Notes on the *Berardius* of New Zealand. *Annals and Magazine of Natural History* 48(8): 115–117.

———. 1874. Notes on Dr. Hector's paper on the whales and dolphins of the New Zealand seas. *Transactions of the New Zealand Institute* 6: 93–97.

Gray, R. W. 1941. The bottlenose whale. *Naturalist* 791: 129–132.

Guiler, E. R. 1966. A stranding of *Mesoplodon densirostris* in Tasmania. *Journal of Mammalogy* 47(2): 327.

———. 1967. Strandings of three species of *Mesoplodon* in Tasmania. *Journal of Mammalogy* 48(4): 650–652.

Gunter, G. 1955. Blainville's beaked whale, *Mesoplodon densirostris*, on the Texas coast. *Journal of Mammalogy* 36(4): 573–574.

Haast, J. Von. 1870. Preliminary notice of a ziphid whale, probably *Berardius arnouxii*, stranded on the 16th December, 1868, on the beach near New Brighton, Canterbury. *Transactions of the New Zealand Institute* 2: 190–192.

Hale, H. M. 1931a. Beaked whales—*Hyperoodon planifrons* and *Mesoplodon layardi*—from South Australia. *Records of the South Australian Museum* 4(3): 291–311.
———. 1931b. The goose-beaked whale (*Ziphius cavirostris*) in New Ireland. *Records of the South Australian Museum* 4(3): 312–313.
———. 1932. The New Zealand scamperdown whale (*Mesoplodon grayi*) in South Australian waters. *Records of the South Australian Museum* 4(4): 489–496.
———. 1939. Rare whales in South Australian waters. *South Australian Naturalist* 19(4): 5–8.
———. 1962. Occurrence of the whale *Berardius arnouxi* in Southern Australia. *Records of the South Australian Museum* 14(2): 231–243.
Hamilton, R. 1839. Mammalia: Whales, etc. In William Jardine, ed., *The Naturalist's Library.* W. H. Lizars, London.
Hanlon, R. T., and J. B. Messenger. 1996. *Cephalopod Behaviour.* Cambridge University Press.
Hardy, M. D., C. D. MacLeod, and J. C. Goold. 2006. Weapon reinforcement: The extent and development of sexually and age dimorphic structures in the skull of the northern bottlenose whale, *Hyperoodon ampullatus.* Poster NH2, 20th Annual Conference of the European Cetacean Society. Gdynia, Poland, April 1–6, 2006.
Harmer, S. F. 1924. On *Mesoplodon* and other beaked whales. *Proceedings of the Zoological Society of London* 1924: 541–587.
———. 1927. Report on cetaceans stranded on the British coasts from 1913 to 1926. *Bulletin of the British Museum (Natural History)* 10: 1–91.
Hass, H. 1959. *We Come from the Sea.* Doubleday.
Hay, K. A., and A. W. Mansfield. 1989. Narwhal *Monodon monoceros.* In S. H. Ridgway and R. Harrison, eds., *Handbook of Marine Mammals. Vol. 4: River Dolphins and the Larger Toothed Whales,* pp. 145–176. Academic Press.
Hector, J. 1873. On the whales and dolphins of the New Zealand seas. *Transactions of the New Zealand Institute* 5: 154–170.
———. 1878. Notes on the whales of the New Zealand seas. *Transactions of the New Zealand Institute* 10: 331–343.
Henshaw, M. D., R. G. Leduc, S. J. Chivers, and A. E. Dizon. 1997. Identification of beaked whales (family Ziphiidae) using mtDNA sequences. *Marine Mammal Science* 13: 487–495.
Hershkovitz, P. 1966. *Catalog of Living Whales. United States National Museum Bulletin* 246. Smithsonian Institution, Washington, DC. 259 pp.
Heyning, J. E. 1984. Functional morphology involved in intraspecific fighting of the beaked whale *Mesoplodon carlhubbsi. Canadian Journal of Zoology* 62: 1645–1654.
———. 1989a. Comparative facial anatomy of beaked whales (Ziphiidae) and a systematic revision among the families of extant Odontoceti. *Contributions in Science* 405: 1–66.
———. 1989b. Cuvier's Beaked Whale—*Ziphius cavirostris.* In S. H. Ridgway and R. Harrison, eds., *Handbook of Marine Mammals. Vol. 4: River Dolphins and the Larger Toothed Whales,* pp. 289–308. Academic Press.
Heyning. J. E., and J. G. Mead. 1996. Suction feeding in beaked whales: Morphological and observational evidence. *Contributions in Science* 464(12): 1–12.
———. 2009. Cuvier's beaked whale *Ziphius cavirostris.* In W. Perrin, B. Würsig, and J. G. M. Thewissen, eds. *Encyclopedia of Marine Mammals,* pp. 294–295. Academic Press.
Holthuis, L. B. 1987. The scientific name of the sperm whale. *Marine Mammal Science* 3(1): 87–88.

Hooker, S. K., and R. W. Baird. 1999a. Deep-diving behaviour of the northern bottlenose whale, *Hyperoodon ampullatus* (Cetacea: Ziphiidae). *Proceedings of Royal Society of London* 266: 671–676.

———. 1999b. Observations of Sowerby's beaked whales, *Mesoplodon bidens*, in the Gully, Nova Scotia. *Canadian Field-Naturalist* 113(2): 273–277.

Hooker, S. K., R. W. Baird, and A. Fahlman. 2009. Could beaked whales get the bends? Effect of diving behaviour and physiology on modelled gas exchange for three species: *Ziphius cavirostris, Mesoplodon densirostris* and *Hyperoodon ampullatus. Respiratory Physiology and Neurobiology* 167: 235–246.

Houck, W. J. 1958. Cuvier's beaked whale from northern California. *Journal of Mammalogy* 39(2): 308–309.

Hubbs, C. L. 1946. First records of two beaked whales *Mesoplodon bowdoini* and *Ziphius cavirostris* from the Pacific coast of the United States. *Journal of Mammalogy* 27(3): 242–255.

Husson, A. M., and L. B. Holthius. 1974. *Physeter macrocephalus* Linnaeus, 1758, the valid name for the sperm whale. *Zoologische Mededelingen* 48(19): 205–217.

International Whaling Statistics. 1978. Sandefjord, Norway. Pp. 1–55.

Jacques, T. G., and R. H. Lambertsen. 1997. *The North Sea Sperm Whales, One Year After.* Institut Royal des Sciences Naturelles de Belgique (Bruxelles).

Jasny, M., J. Reynolds, C. Horowitz, and A. Wetzler. 2005. *Sounding the Depths II: The Rising Toll of Sonar, Shipping, and Industrial Noise on Marine Life.* National Resources Defense Council.

Jefferson, T. A., and G. D. Baumgardner. 1997. Osteological specimens of marine mammals (Cetacea and Sirenia) from the Western Gulf of Mexico. *Texas Journal of Science* 49(2): 97–108.

Jefferson, T. A., D. Fertl, M. Michael, and T. D. Fagin. 2006. An unusual encounter with a mixed school of melon-headed whales (*Peponocephala electra*) and rough-toothed dolphins (*Steno bredanensis*) at Rota, Northern Mariana Islands. *Micronesia* 38(2): 239–244.

Jefferson, T. A., M. A. Webber, and R. L. Pitman. 2008. *Marine Mammals of the World: A Comprehensive Guide to their Identification.* Academic Press.

Jellison, W. M. 1953. A beaked whale, *Mesoplodon* sp., from the Pribilofs. *Journal of Mammalogy* 34(2): 249–251.

Jepson, P. D., M. Arbelo, R. Deaville, I. A. P. Patterson, P. Castro, J. R. Baker, E. Degollada, H. M. Ross, P. Herráez, A. M. Pocknell, F. Rodríguez, F. E. Howie, A. Espinosa, R. J. Reid, J. R. Jaber, V. Martin, A. A. Cunningham, and A. Fernández. 2003. Gas bubble lesions in stranded cetaceans. *Nature* 425: 575–576.

Jepson, P. D., R. Deaville, K. Acevedo-Whitehouse, J. Barnett, A. Brownlow, R. L. Brownell, F. C. Clare, N. Davison, R. J. Law, J. Loveridge, S. K. Macgregor, S. Morris, S. Murphy, R. Penrose, M. W. Perkins, E. Pinn, H. Seibel, U. Siebert, E. Sierra, V. Simpson, M. L. Tasker, N. Tregenza, A. A. Cunningham, A. Fernández. 2013. What caused the UK's largest common dolphin (*Delphinus delphis*) mass stranding event? *PLoS One* 8(4): e60953. doi:10.371/jpouirnal.pone.0060953.

Johnson G., A. Frantzis, C. Johnson, V. Alexiadou, S. Ridgway, and P. T. Madsen 2010. Evidence that sperm whale calves (*Physeter macrocephalus*) suckle through their mouth. *Marine Mammal Science* 26(4): 990–996.

Johnson, M. P., L. S. Hickmott, N. Aguilar de Soto, and P. T. Madsen. 2008. Echolocation behavior adapted to prey in foraging Blainville's beaked whale (*Mesoplodon densirostris*). *Proceedings of the Royal Society of London B: Biological Sciences* 275: 133–139.

Johnson, M. P., P. Madsen, W. M. X. Zimmer, N. Aguilar de Soto, and P. L. Tyack. 2004. Beaked whales echolocate on prey. *Proceedings of the Royal Society of London B: Biological Sciences* 271: S383–S386.

———. 2006. Foraging Blainville's beaked whales (*Mesoplodon densirostris*) produce different distinct click types matched to different phases of echolocation. *Journal of Experimental Biology* 209: 5038–5050.

Johnston, D. W., M. McDonald, J. Polovina, R. Domokos, S. Wiggins, and J. Hildebrand. 2008. Temporal patterns in the acoustic signals of beaked whales at Cross Seamount. *Biology Letters* 4: 208–211.

Jones, B. A., K. Stanton. A. C. Lavery, M. P. Johnson, P. T. Madsen, and P. L. Tyack. 2008. Classification of broadband echoes from prey of a foraging Blainville's beaked whale. *Journal of the Acoustical Society of America* 123(3): 1753–1762.

Jones, E. C. 1971. *Isistius brasiliensis*, a squaloid shark, the probable cause of crater wounds on fishes and cetaceans. *Fishery Bulletin* 69(4): 791–798.

Jonsgard, A. 1955. Development of the modern Norwegian small whale industry. *Norsk Hvalfangst-Tidende* 57(3): 164–167.

———. 1977. A note on the value of bottlenose whales in relation to minke whales and the influence of the market situation and the prices on Norwegian whaling activity. *Reports of the International Whaling Commission* 27: 502–504.

Jonsgard, A., and P. Hoidal. 1957. Strandings of Sowerby's whale (*Mesoplodon bidens*) on the west coast of Norway. *Norsk Hvalfangst-Tidende* 46(12): 507–512.

Julian, F., and M. Beeson. 1997. Estimates of marine mammal, turtle, and seabird mortality for two California gillnet fisheries: 1990–1995. *Fishery Bulletin* 96: 271–284.

Kasuya, T. 1971. Consideration of distribution and migration of toothed whales off the Pacific coast of Japan based upon aerial sighting record. *Scientific Reports of the Whales Research Institute* 32: 37–60.

———. 1977. Age determination and growth of the Baird's beaked whale with a comment on the fetal growth rate. *Scientific Reports of the Whales Research Institute* 29: 1–20.

———. 1988. Distribution and Behavior of Baird's beaked whales off the Pacific coast of Japan. *Scientific Reports of the International Whaling Commission* SC/40/SM11: 1–20.

Kasuya, T., R. L. Brownell, and K. C. Balcomb. 1988. Preliminary analysis of life history of Baird's beaked whale off the Pacific coast of central Japan. *Scientific Reports of the International Whaling Commission* SC/40/SM18: 1–21.

Kasuya, T., and M. Nishiwaki. 1971. First record of *Mesoplodon densirostris* from Formosa. *Scientific Reports of the Whales Research Institute* 23: 129–137.

Kasuya, T., and S. Ohsumi. 1988. Supplemental analysis of the Baird's beaked whale stock in the western North Pacific. *Scientific Reports of the International Whaling Commission* SC/40/SM25: 1–19.

Kellogg, R. 1928. The history of whales—their adaptation to life in the water. *Quarterly Review of Biology* 3(1): 29–76, 174–208.

Kenney, R. D., M. A. M. Hyman, R. E Owen, G. P. Scott, and H. E. Winn. 1986. Estimation of prey densities required by western North Atlantic right whales. *Marine Mammal Science* 2(1): 1–13.

Kenney, R. D. 2009. Right whales *(Eubalaena glacialis, E. japonica* and *E. australis).* In W. F. Perrin, B. Würsig and J. G. M. Thewissen, eds., *Encyclopedia of Marine Mammals,* pp. 963–972. Academic Press.

Kenyon, K. W. 1961. Cuvier beaked whales stranded in the Aleutian Islands. *Journal of Mammalogy* 42(1): 71–76.

Kernan, J. D. 19 18. The skull of *Ziphius cavirostris. Bulletin of the American Museum of Natural History* 38: 349–394.

Kiekow, T. 2009. Sonar—a severe blow to beaked whales. *Whalewatcher* 38(1): 23–25.

Kitamura, S., T. Matsuishi, T. K. Yamada, Y. Tajima, H. Ishikawa, S. Tanabe, H. Nakagawa, Y. Uni, and S. Abe. 2013. Two genetically distinct stocks in Baird's beaked whale (Cetacea: Ziphiidae). *Marine Mammal Science* 29(4): 755–766.

Kovacic, I., M. D. Gomercic, H. Gomercic, H. Lucic, and T. Gomercic. 2010. Stomach contents of two Cuvier's beaked whales (*Ziphius cavirostris*) stranded in the Adriatic Sea. *Marine Biodiversity Records* 3: 1–4.

Krefft, C., and J. E. Gray. 1871. Notice of a new Australian ziphioid whale. *Annals and Magazine of Natural Hist*ory 7: 368.

Kyhn, L. A, J. Tougaard, K. Beedholm, F. H. Jensen, E. Ashe, R. Williams, and P. T. Madsen. 2013. Clicking in a killer whale habitat: Narrow-band, high-frequency biosonar clicks of harbour porpoise (*Phocoena phocoena*) and Dall's porpoise (*Phocoenoides dalli*). *PLoS ONE* 8(5): e63763. doi: 10.1371/ journal.pone.0063763.

Lambert, O. 2005. Systematics and phylogeny of the fossil beaked whales *Ziphirostrum* du Bus 1868 and *Choneziphius* Duvernoy 1851 (Cetacea, Odontoceti) from the Neogene of Antwerp (North of Belgium). *Geodiversitas* 27(1): 45430–45497.

Lambert, O., G. Bianucci, and K. Post. 2009. A new beaked whale (Odontoceti, Ziphiidae) from the Middle Miocene of Peru. *Journal of Vertebrate Paleontology* 29(3): 910–922.

———. 2010. Tusk-bearing beaked whales from the Miocene of Peru: Sexual dimorphism in fossil ziphiids? *Journal of Mammalogy* 91(1): 19–26.

Lambert, O., V. de Buffrénil, and C. de Muizon. 2011. Rostral densification in beaked whales: Diverse processes for a similar pattern. *Comtes Rendus Palevol.* 120: 453–468.

Lambert, O., C. de Muizon, and G. Bianucci. 2013. The most basal beaked whale, *Ninoziphius platyrostris* Muizon, 1983: Clues on the evolutionary history of the family Ziphiidae (Cetacea: Odontoceti). *Zoological Journal of the Linnean Society* 167(4): 569–598.

Lambert, O., S. J. Godfrey, and A. J. Fuller. 2010. A Miocene ziphiid (Cetacea: Odontoceti) from Calvert Cliffs, Maryland, U.S.A. *Journal of Vertebrate Paleontology* 30: 1645–1651.

Lambert, O., and S. Louwye. 2006. *Archeoziphius microglenoides,* a new primitive beaked whale from the Middle Miocene of Belgium. *Journal of Vertebrate Paleontology* 26: 910–922.

Laporta, P., R. Praderi, V. Little, and A. Le Bas. 2005. An Andrew's beaked whale *Mesoplodon bowdoini* (Cetacea, Ziphiidae) stranded on the Atlantic coast of Uruguay. *Latin American Journal of Aquatic Mammals* 4(2): 101–111.

Leatherwood, S., and R. R. Reeves. 1983. *The Sierra Club Handbook of Whales and Dolphins.* Sierra Club Books.

Li, Z., and J. D. Pasteris 2014a. Chemistry of bone mineral, based on the hypermineralized rostrum of the beaked whale *Mesoplodon densirostris. American Mineralogist* 99: 645–653.

———. 2014b. Tracing the pathway of compositional changes in bone mineral with age: Preliminary study of bioapatite aging in hypermineralized dolphin's bulla. *Biochimica et Biophysica Acta* 1840 (2014): 2331–2339.

Lin, A. 2013. Anatomical description and phylogenetic analysis of Miocene beaked whale from the East African Rift Valley, Kenya. *Engaged Learning Projects Journal*, paper 10. http//:digital repository.smu.edu/upjournal_research/10.

Linnaeus, C. 1758. *Systema naturae per regna tria naturae, secundum classes, ordines, genera, species, cum characteribus, differentiis, synonymis, locis*. Laurentii Salvii, Holmiae, editio decima, reformata, tomus i.

Liouville, J. 1913. *Cetacés de l'Antarctique (Baleinopteres, Ziphiides, Delphinides): Deuxième Expedition Antarctique Français, 1908–1919*. Masson, Paris.

Longman, H. A. 1926. New records of Cetacea with a list of Queensland species. *Memoirs of the Queensland Museum* 8(3): 266–278.

Loughlin, T. R., and M. A. Perez. 1985. *Mesoplodon stejnegeri*. *Mammalian Species* 250: 1–6.

Lynne, S. K., and D. L. Reiss. 1992. Pulse sequence and whistle production by two captive beaked whales, *Mesoplodon* species. *Marine Mammal Science* 8(3): 299–305.

Mackintosh, N. A., and J. F. G. Wheeler. 1929. *Southern Blue and Fin Whales*. Discovery Reports I: 257–540, Plates XXV–XLIV.

MacLeod, C. D. 1998. Intraspecific scarring in odontocete cetaceans: An indicator of male "quality" in aggressive social interactions. *Journal of the Zoological Society of London* 244: 71–77.

———. 2000. Species recognition as a possible function for variations in position and shape of the sexually dimorphic tusks of *Mesoplodon* whales. *Evolution* 54(6): 2171–2173.

———. 2002. Possible functions of the ultradense bone in the rostrum of Blainville's beaked whale (*Mesoplodon densirostris*). *Canadian Journal of Zoology* 80: 178–184.

———. 2006. How big is a beaked whale? A review of body length and sexual size dimorphism in the family Ziphiidae. *Journal of Cetacean Research and Management* 7(3): 301–308.

MacLeod, C. D., and A. D'Amico. 2006. A review of beaked whale behaviour and ecology in relation to assessing and mitigating impacts of anthropogenic noise. *Journal of Cetacean Research and Management* 7(3): 211–221.

MacLeod, C. D., M. D. Hardy, and J. C. Goold. 2006. Adaptations and evolution of structures for intraspecific combat in the family Ziphiidae. Poster GE5, 20th Annual Conference of the European Cetacean Society. Gdynia, Poland, April 1–6, 2006.

MacLeod, C. D., and J. S. Herman. 2004. Development of tusks and associated structures in *Mesoplodon bidens* (Cetaceae, Mammalia). *Mammalia* 68(2–3): 175–184.

MacLeod, C. D., W. F. Perrin, R. T. Pitman, J. Barlow, L. Ballance, A. D'Amico, T. Gerodette, G. Joyce, K. D. Mullin, D. L. Palka, and G. T. Waring. 2006. Known and inferred distributions of beaked whale species (Cetacea: Ziphiidae). *Journal of Cetacean Research and Management* 7(3): 271–286.

MacLeod, C. D., J. S. Reidenberg, M. Weller, M. B. Santos, J. Herman, J. Goold, and G. J. Pierce. 2009. Breaking symmetry: The marine environment, prey size, and the evolution of asymmetry in cetacean skulls. *Anatomical Record* 290: 539–545.

MacLeod, C. D., M. B. Santos, and G. J. Pierce. 2003. Review of data on diets of beaked whales: Evidence of niche separation and geographic separation. *Journal of the Marine Biological Association of the U.K.* 83: 651–665.

MacLeod, C. D., M. B. Santos, A López, and G. J. Pierce. 2006. Relative prey size consumption in toothed whales: Implications for prey selection and level of specialization. *Marine Ecology Progress Series* 326: 295–307.

MacLeod, C. D., and A. F. Zuur. 2005. Habitat utilization by Blainville's beaked whales off Great Abaco, northern Bahamas, in relation to seabed topography. *Marine Biology* 147: 1–11.

Madsen, P. T. 2012. Foraging with the biggest nose on record. *Whalewatcher* 41(1): 9–15.

Madsen, P. T., N. Aguilar de Soto, P. Arranz, and M. Johnson. 2013. Echolocation in Blainville's beaked whales (*Mesoplodon densirostris*). *Journal of Comparative Physiology* 199: 451–469.

Madsen, P. T., D. A. Carder, W. L. Au, P. E. Nachtigall, B. Møhl, and S. H. Ridgway. 2003. Sound production in neonate sperm whales. *Journal of the Acoustical Society of America* 113: 2988–2991.

Madsen, P. T., M. Johnson, N. Aguilar de Soto, W. M. X. Zimmer, and P. Tyack. 2005. Biosonar performance of foraging beaked whales (*Mesoplodon densirostris*). *Journal of Experimental Biology* 208: 181–194.

Madsen, P. T., R. Payne, U. Kristiansen, M. Wahlberg, I. Kerr, and B. Møhl. 2002. Sperm whale sound production studied with ultrasound time/depth recording tags. *Journal of Experimental Biology* 205: 1899–1906.

Madsen, P. T., and A. Surlykke. 2013. Functional convergence in bat and toothed whale biosonars. *Physiology* 28: 276–283.

Madsen, P. T., M. Wahlberg, and B. Møhl. 2002. Male sperm whale (*Physeter macrocephalus*) acoustics in a high latitude: Implications for echolocation and communication. *Behavioral Ecology and Sociobiology* 53: 31–41.

Madsen, P. T., M. Wilson, M. Johnson, R. T. Hanlon, A. Bocconcelli, N. Aguilar de Soto, and P. L. Tyack. 2007. Clicking for calamari: Toothed whales can echolocate squid *Loligo pealeii*. *Aquatic Biology* 1(2): 141–150.

Malakoff, D. 2002. Seismology: Suit ties whale deaths to research cruise. *Science* 298: 722–723.

Manire, C. A., H. L. Rhinehart, N. B. Barros, L. Byrd, and P. Cunningham-Smith. 2004. An approach to the rehabilitation of *Kogia* spp. *Aquatic Mammals* 30(2): 257–270.

Marino, L., D. Sol, K. Toren, and L. Lefebvre. 2006. Does diving limit brain size in cetaceans? *Marine Mammal Science* 22(2): 413–425.

Marlow, B. J. 1963. Rare beaked whale washed up on Sydney beach (*Mesoplodon layardi*). *Australian Natural History* 14: 164.

Martin, V., M. Tejedor, M. Pérez-Gil, M. L. Dalebout, M. Arbelo, and A. Fernández. 2011. A Sowerby's beaked whale (*Mesoplodon bidens*) stranded in the Canary Islands: The most southern record in the eastern North Atlantic. *Aquatic Mammals* 37(4): 512–519.

Matsumoto, H., C. Jones, H. Klinck, D. K. Mellinger, R. P. Dziak, and C. Meinig. 2013. Tracking beaked whales with a passive acoustic profiler float. *Journal of the Acoustical Society of America* 133: 731.

Matthiessen, P. 1959. *Wildlife in America*. Viking Press.

Maturana, H. R., and S. Sperling. 1963. Unidirectional response to angular acceleration recorded from the middle cristal nerve in the statocyst of *Octopus vulgaris*. *Nature* 197: 815–816.

Mayr, E. 1963. *Animal Species and Evolution.* Harvard University Press.

Mcalpine, D. F. 2009. Pygmy and dwarf sperm whales. In W. F. Perrin, B. Würsig, and J. G. M. Thewissen, eds., *Encyclopedia of Marine Mammals,* pp. 936–938. Academic Press.

McCann, C. 1961. The occurrence of the southern bottlenosed whale, *Hyperoodon planifrons* Flower, in New Zealand waters. *Records of the Dominion Museum of Wellington* 4(3): 21–27.

———. 1962a. The taxonomic status of the beaked whale, *Mesoplodon pacificus* Longman (Cetacea). *Records of the Dominion Museum of Wellington* 4: 95–100.

———. 1962b. The taxonomic status of the beaked whale *Mesoplodon hectori* (Gray) Cetacea. *Records of the Dominion Museum of Wellington* 4: 83–94.

———. 1962c. Key to the Family Ziphiidae (beaked whales). *Tuatara* 10(1): 13–18.

———. 1964. The female reproductive organs of Layard's beaked whale, *Mesoplodon layardi* (Gray). *Records of the Dominion Museum of Wellington* 4: 311–316.

———. 1974. Body scarring on cetacea—odontocetes. *Scientific Reports of the Whales Research Institute* 26: 145–155.

———. 1975. A study of the genus *Berardius* Duvernoy. *Scientific Reports of the Whales Research Institute* 27: 111–137.

McCann, C., and F. H. Talbot. 1963. The occurrence of True's beaked whale (*Mesoplodon mirus* True) in South African waters, with a key to South African species of the genus. *Proceedings of the Linnaean Society of London* 175(2): 137–145.

McCarthy, W. J. 1948. The uses of ASDICS in whaling. *Polar Record* 5: 220–232.

McCormick, J. G., E. G. Wever, S. H. Ridgway, and J. Palin. 1979. Sound reception in the porpoise as it is related to echolocation. In R.-G. Busnel and J. G. Fish, eds., *Animal Sonar Systems,* pp. 449–467. Plenum.

McDonald, M. A.. J. A. Hildebrand, S. M. Wiggins, D. W. Johnson, and J. J. Polovina 2009. An acoustic survey of beaked whales at Cross Seamount near Hawaii. *Journal of the Acoustical Society of America* 125(1): 624–627.

McLachlan, G. R., R. Liversidge, and R. M. Tietz. 1966. A record of *Berardius arnouxi* from the south-east coast of South Africa. *Annals of the Cape Province Museum (Natural History)* 5: 91–100.

McSweeney, D. J., R. W. Baird, and S. D. Mahaffy. 2007. Site fidelity, associations, and movements of Cuvier's (*Ziphius cavirostris*) and Blainville's (*Mesoplodon densirostris*) beaked whales off the island of Hawai'i. *Marine Mammal Science* 23: 666–687.

Mead, J. G. 1975. A fossil beaked whale (Cetacea: Ziphiidae) from the Miocene of Kenya. *Journal of Paleontology* 49(4): 745–751.

———. 1981. First records of *Mesoplodon hectori* (Ziphiidae) from the northern hemisphere and a description of the adult male. *Journal of Mammalogy* 62(2): 430–432.

———. 1984. Survey of reproductive data for beaked whales (Ziphiidae). *Report of the International Whaling Commission* Special Issue 6: 91–95.

———. 1989a. Shepherd's Beaked Whale—*Tasmacetus shepherdi.* In S. H. Ridgway and R. Harrison, eds., *Handbook of Marine Mammals. Vol. 4: River Dolphins and the Larger Toothed Whales,* pp. 309–320. Academic Press.

———. 1989b. Bottlenose Whales. *Hyperoodon ampullatus* (Forster, 1770) and *Hyperoodon planifrons* Flower, 1882. In S. H. Ridgway and R. Harrison, eds., *Handbook of Marine Mammals. Vol. 4: River Dolphins and the Larger Toothed Whales,* pp. 321–348. Academic Press.

———. 1989c. Beaked Whales of the Genus *Mesoplodon*. In S. H. Ridgway and R. Harrison, eds., *Handbook of Marine Mammals. Vol. 4: River Dolphins and the Larger Toothed Whales*, pp. 349–430. Academic Press.

———. 2007. Stomach anatomy and use in defining systemic relationships of the cetacean family Ziphiidae (beaked whales). *Anatomical Record* 290: 581–595.

———. 2009. Beaked whales, overview. *Ziphiidae*. In W. Perrin, B. Würsig, and J. G. M. Thewissen, eds., *Encyclopedia of Marine Mammals*, pp. 94–97. Academic Press.

Mead, J. G., and A. N. Baker. 1987. Notes on the rare beaked whale *Mesoplodon hectori* (Gray). *Journal of the Royal Society of New Zealand* 17: 303–312.

Mead, J. G., and R. L. J. Brownell. 2005. "Order Cetacea." In D. E. Wilson and D. M. Reeder, *Mammal Species of the World*, third edition, pp. 723–743. Johns Hopkins University Press, Baltimore.

Mead, J. G., and R. Payne. 1975. A specimen of the Tasman beaked whale (*Tasmacetus shepherdi*) from Argentina. *Journal of Mammalogy* 56(1): 213–218.

Mead, J. J., W. A. Walker, and W. J. Houck. 1982. Biological observations on *Mesoplodon carlhubbsi* (Cetacea: Ziphiidae). *Smithsonian Contributions to Zoology* 344: 25 pp.

Melville, H. 1851. *Moby-Dick; or the Whale.* Harper and Bros., New York.

Melville, H. 1851. *The whale.* Richard Bentley, London.

Melville, R. V. 1995. *Towards Stability in the Names of Animals—a History of the International Commission on Zoological Nomenclature 1895–1995.* International Trust for Zoological Nomenclature, London.

Miller, P. J. O., M. P. Johnson, and P. L. Tyack. 2004. Sperm whale behaviour indicates the use of echolocation click buzzes 'creaks' in prey capture. *Proceedings of the Royal Society B* 271: 2239–2247.

Milne, A. R. 1959. Comparison of spectra of an earthquake T-phase with similar signals from nuclear explosions. *Bulletin of the Seismological Society of America* 49(4): 317–329.

Miralles, L., S. Lens, A. Rodríguez-Folgar, M. Carrillo, V. Martín, B. Mikkelsen, and E. Garcia-Vazquez. 2013. Interspecific introgression in cetaceans: DNA markers reveal post-F1 status of a pilot whale. *PLoS ONE* 8(8): e69511.

Mitchell, E. D. 1968. Northeast Pacific stranding distribution and seasonality of Cuvier's beaked whale *Ziphius cavirostris*. *Canadian Journal of Zoology* 46: 265–279.

———. 1975. *Porpoise, Dolphin and Small Whale Fisheries of the World.* IUCN, Morges, Switzerland.

———. 1977. Evidence that the northern bottlenose whale is depleted. *Report of the International Whaling Commission* 27: 195–205.

Mitchell, E. D., and W. J. Houck. 1967. Cuvier's beaked whale (*Ziphius cavirostris*) stranded in northern California. *Journal of the Fisheries Research Board of Canada* 24(2): 2503–2513.

Mitchell, E. D., and V. M. Kozicki. 1975. Autumn stranding of a northern bottlenose whale (*Hyperoodon ampullatus*) in the Bay of Fundy, Nova Scotia. *Journal of the Fisheries Research Board of Canada* 37(7): 1019–1040.

Miyashita, T. 1988. Abundance of Baird's beaked whales off the Pacific coast of Japan and possible bias of estimate. *Scientific Reports of the International Whaling Commission* SC/40/SM20: 1–39.

Miyazaki, N., and S. Wada. 1978. Observation of cetacea during whale marking cruise in the western tropical Pacific, 1976. *Report of the International Whaling Commission* 30: 179–195.

Montgomery, S. H., J. H. Geisler, M. R. McGowen, C. Fox, L. Marino, and J. Gatesy. 2013. The evolutionary history of cetacean brain and body size. *Evolution.* doi.10.111.evo.12197.

Mooney, T. A., R. T. Hanlon, P. T. Madsen, J. Christensen-Dalsgaard, D. R. Ketten, and P. E. Nachtigall. 2012. Sound detection by the longfin squid (*Loligo pealeii*) studied with auditory evoked potentials: Sensitivity to low frequency particle motion and not pressure. *Journal of Experimental Biology* 213: 3748–3759.

Moore, J. C. 1958. A beaked whale from the Bahama Islands and comments on the distribution of *Mesoplodon densirostris. American Museum Novitates* 1897: 1–12.

———. 1960. New records of the Gulf-Stream beaked whale, *Mesoplodon gervaisi,* and some taxonomic considerations. *American Museum Novitates* 1993: 1–35.

———. 1963. Recognizing certain species of beaked whales of the Pacific Ocean. *American Midland Naturalist* 70(2): 396–428.

———. 1966. Diagnoses and distributions of beaked whales of the genus *Mesoplodon* known from North American waters. In K. S. Norris, ed., *Whales, Dolphins and Porpoises,* pp. 32–61. University of California Press.

———. 1968. Relationships among the living genera of beaked whales with classifications, diagnoses and keys. *Fieldiana*: *Zoology* 53(4): 209–298.

———. 1972. More skull characters of the beaked whale *Indopacetus pacificus* and comparative measurements of austral relatives. *Fieldiana: Zoology* 62: 1–19.

Moore, J. E., and J. P. Barlow. 2013. Declining abundance of beaked whales (Family Ziphiidae) in the California Current Large Ecosystem. *PLoS ONE* 8(1): e52770. doi:10.1371/journal .pone.0052770.

Moore, J. C., and R. M. Gilmore. 1965. A beaked whale new to the Western Hemisphere. *Nature* 205: 1239–1240.

Moore, J. C., and F. G. Wood. 1957. Differences between the beaked whales *Mesoplodon mirus* and *Mesoplodon gervaisi. American Museum Novitates* 1831: 1–25.

Mörzer Bruyns, W. F. J. 1971. *Field guide of whales and dolphins.,* N. V. Uitgeverij, Amsterdam.

Moynihan, M. 1985a. *Communication and Noncommunication by Cephalopods.* Indiana University Press.

———. 1985b. Why are cephalopods deaf? *American Naturalist* 125(3): 465–469.

Muizon, C. de. 1983. Un Ziphiidae (Cetacea) nuveau Pliocene inferieur du Perou. *Comptes rendus de l'Academie des sciences de Paris* 297: 85–88.

———. 1991. A new Ziphiidae (Cetacea) from the early Miocene of Washington state (USA) and phylogenetic analysis of the major groups of odontocetes. *Bulletin du Musee National d'Histoire Naturelle, Paris,* 4e serie, section C 12(3–4): 279–326.

———. 1993a. *Odobenocetops peruvianus*: Una remarcable convergencia de adaptación alimentaria entre morse y delfín. *Bull. Inst. Français d'Études Andines* 2(3): 671–683.

———. 1993b. Walrus-like feeding adaptation in a new cetacean from the Pliocene of Peru. *Nature* 365: 745–748.

Muizon, C. de., D. P. Domning, and M. Parish. 1999. Dimorphic tusks and adaptive strategies in a new species of walrus-like dolphin (Odobenocetopsidae) from the Pliocene of Peru. *Comptes Rendus Academie des Sciences; Sci. Terre Planètes* 329: 449–455.

Nansen, F. 1925. *Hunting and Adventure in the Arctic*. Duffield.
Ness, A. R. 1967. A measure of asymmetry in skulls of odontocete whales. *Journal of Zoology* 153: 209i–221.
Nicol, J. 2013. Whale Meat at $22 a Kilo. *Marlborough Express*. October 5.
Nilsson, D.-E., E. J. Warrant, S. Johnsen, R. Hanlon, and N. Shashar. 2012. A unique advantage for giant eyes in giant squid. *Current Biology* 22: 1–6.
Nishimura, S., and M. Nishiwaki. 1964. Records of the beaked whale *Mesoplodon* from the Japan Sea. *Publications of the Seto Marine Biological Laboratory* 12(4): 323–334.
Nishiwaki, M. 1962a. *Mesoplodon bowdoini* stranded at Akita Beach, Sea of Japan. *Scientific Report of the Whales Research Institute* 16: 61–77.
———. 1962b. Observations on two mandibles of *Mesoplodon*. *Scientific Report of the Whales Research Institute* 16: 79–82.
———. 1972. *Mammals of the Sea: Biology and Medicine*. Charles C. Thomas. 204 pp.
Nishiwaki, M., and T. Kamiya. 1958. A beaked whale *Mesoplodon* stranded at Oiso Beach, Japan. *Scientific Report of the Whales Research Institute* 13: 53–83.
———. 1959. *Mesoplodon stejnegeri* from the coast of Japan. *Scientific Report of the Whales Research Institute* 14: 35–48.
Nishiwaki, M., T. Kasuya, K. Kurehara, and N. Oguro. 1972. Further comments on *Mesoplodon ginkgodens*. *Scientific Report of the Whales Research Institute* 24: 43–56.
Nishiwaki, M., and N. Oguro. 1971. Baird's beaked whales caught on the coast of Japan in recent 10 years. *Scientific Report of the Whales Research Institute* 23: 111–122.
———. 1972. Catch of the Cuvier's beaked whale off Japan in recent years. *Scientific Report of the Whales Research Institute* 25: 33–41.
Norman, J. R., and F. G. Fraser. 1938. *Giant Fishes, Whales and Dolphins*. W. W. Norton.
Norman, S. A., and J. G. Mead. 2001. *Mesoplodon europaeus*. *Mammalian Species* 688: 1–5.
Norris, K. S. 1964. Some problems of echolocation in cetaceans. In W. N. Tavolga, ed., *Marine Bioacoustics*, pp. 317–336. Pergamon Press.
———. 1974. *The porpoise watcher*. Norton, New York.
Norris, K. S., and G. W. Harvey. 1972. A theory for the function of the spermaceti organ in the sperm whale (*Physeter catodon* L.). In S. R. Galler, K. Schmidt-Koenig, G. J. Jacobs, and R. E. Belleville, eds., *Animal Orientation and Navigation*, pp. 397–419. NASA.
Norris, K. S., and B. Møhl. 1983. Can odontocetes debilitate prey with sound? *American Naturalist* 122(1): 85–104.
Norris, K. S., and J. H. Prescott. 1961. Observations on Pacific cetaceans of Californian and Mexican waters. *University of California Publications in Zoology* 63(4): 291–402.
Nweeia, M. T., F. C. Eichmiller, P. V. Hauschka, G. A. Donahue, J. R. Orr, S. H. Ferguson, C. A. Watt, J. G. Mead, C. W. Potter, R. Dietz, A. A. Giuseppetti, S. R. Black, A. J. Trachtenberg, and W. P. Kuo. 2014. Sensory ability in the narwhal tooth organ system. *Anatomical Record* 297: 599–617.
Ohlin, A. 1893. Some remarks on the bottlenose-whale (*Hyperoodon*). *Lunds. Univ. Arskr.* 29: 1–14.
Ohsumi, S. 1975. Review of Japanese small-type whaling. *Journal of the Fisheries Research Board of Canada* 32(7): 1111–1121.
———. 1980. Catches of sperm whales by modern whaling in the North Pacific. *Report of the International Whaling Commission* Special Issue 2: Sperm Whales: 11–19.

Olaus Magnus. 1539. Carta marina et Descriptio septemtrionalium terrarum ac mirabilium rerum in eis contentarum, diligentissime elaborata Anno Domini 1539 Veneciis liberalitate Reverendissimi Domini Ieronimi Quirini [A Marine map and Description of the Northern Lands and of their Marvels, most carefully drawn up at Venice in the year 1539 through the generous assistance of the Most Honourable Lord and Patriarch Hieronymo Quirino]. Rome.

Olaus Magnus. 1555. Historia de gentibus septentrionalibus earumque diversis statibus, conditionibus, moribus, ritibus, superstitionibus . . . [A history of the northern peoples and their diverse social states, conditions, customs, religious practices, superstitions . . .] Viottis, Rome.

Oliver, W. R. B. 1922. The whales and dolphins of New Zealand. *New Zealand Journal of Science and Technology* 5(3): 129–141.

———. 1924. Strap-toothed whale at Kaitawa Point, entrance to Porirua Harbor. *New Zealand Journal of Science and Technology* 7: 187–188.

———. 1937. *Tasmacetus shepherdi*: A new genus and species of beaked whale from New Zealand. *Proceedings of the Zoological Society of London*. Ser. B, 107: 371–381.

Oliviera, C., M. Wahlberg, M. Johnson, P. J. O. Miller, and P. T. Madsen. 2013. The function of male sperm whale slow clicks in high latitude habitat: Communication, echolocation, or prey debilitation? *Journal of the Acoustical Society of America* 133(5): 3135–3144.

Omura, H. 1972. An osteological study of the Cuvier's beaked whale, *Ziphius cavirostris*, in the northwest Pacific. *Scientific Report of the Whales Research Institute* 24: 1–34.

Omura, H., K. Fujino, and S. Kimura. 1955. Beaked whales *Berardius bairdii* off Japan, with notes on *Ziphius cavirostris*. *Scientific Report of the Whales Research Institute* 10: 89–132.

Orr, R. T. 1953. Beaked whale (*Mesoplodon*) from California with comments on taxonomy. *Journal of Mammalogy* 34(2): 239–249.

Osterhaus, A. D., R. L. de Swart, H. W. Vos, P. S. Ross, M. J. Kenter, and T. Barrett. 1995. Morbillivirus infections of aquatic mammals: Newly identified members of the genus. *Veterinary Microbiology* 44(2–4): 219–27.

Packard, A., H. E. Karlsen, and O. Sand 1990. Low frequency hearing in cephalopods. *Journal of Comparative Physiology* 166: 501–505.

Paintadosi, C. A., and E. D. Thalmann. 2004. Pathology: Whales, sonar, and decompression sickness. *Nature* 428: 1.

Palacios, D. M. 1996. On the specimen of the ginkgo-toothed beaked whale, *Mesoplodon ginkgodens*, from the Galápagos Islands. *Marine Mammal Science* 12(3): 444–446.

Paterson, R. A., and S. Van Dyck. 1990. Records of beaked whales in Queensland. *Scientific Reports of Cetacean Research* 1: 63–77.

Perkins, P. J., M. P. Fish, and W. H. Mowbray. 1966. Underwater communication sounds of the sperm whale. *Norsk Hvalfangst-Tidende* 55(12): 225–228.

Perrin, W. F. 2009. The accidental whale. *Whalewatcher* 38(1): 12–15.

Pike, G. C. 1951. Lamprey marks on whales. *Journal of the Fisheries Research Board of Canad*, 8: 275–280.

———. 1953. Two records of *Berardius bairdii* from the coast of British Columbia. *Journal of Mammalogy* 34(1): 98–104.

Pirotta, E., R. Milor, N. Quick, D. Moretti, N. Di Marzio, P. Tyack, I. Boyd, and G. Hastie. 2012. Vessel noise affects beaked whale behavior: Results of a dedicated acoustic response study. *PLoS One* 7(8): e42535. doi: 10.1371/journal.pone.0042535.

Pitman, R. L. 2002. Alive and whale—a missing cetacean resurfaces in the tropics. *Natural History* 2002(9): 32, 34–36.

Pitman, R. L. 2005. Rare whale spotted from icebreaker. *Antarctic Sun,* January 9, p. 4.

———. 2009. Mesoplodont whales (*Mesoplodon* spp.). In W. Perrin, B. Würsig, and J. G. M. Thewissen, eds., pp. 721–726. *Encyclopedia of Marine Mammals.* Academic Press.

Pitman, R. L., A. Aguayo L., and J. Urbán R. 1987. Observations of an unidentified beaked whale (*Mesoplodon* sp.) in the Eastern Tropical Pacific. *Marine Mammal Science* 3(4): 345–352.

Pitman, R. L., and M. S. Lynn. 2001. Biological observations of an unidentified mesoplodont whale in the Eastern Tropical Pacific and probable identity: *Mesoplodon peruvianus. Marine Mammal Science* 17(3): 648–657.

Pitman, R. L., D. M. Palacios, P. L. R. Brennan, B. J. Brennan, K. C. Balcomb, and T. Miyashita. 1999. Sightings and possible identity of a bottlenose whale in the tropical Indopacific: *Indopacetus pacificus? Marine Mammal Science* 15(2): 531–549.

Pitman, R. L., A. L. van Helden, P. B. Best, and A. Pym. 2006. Shepherd's beaked whale (*Tasmacetus shepherdi*): Information on appearance and biology based on strandings and at-sea observations. *Marine Mammal Science* 22(3): 744–755.

Pivorunas, A. 1979. The feeding mechanisms of baleen whales. *American Scientist* 67(4): 432–440.

Podesta, M., L. Cagnolaro, and B. Cozzi. 2005. First record of a stranded Gervais beaked whale, *Mesoplodon europaeus* (Gervais, 1855), in the Mediterranean waters. *Atti della Societa Italiana di Scienze Naturali e del Museo Civico di Storia Naturale in Milano* 146(I): 109–116.

Podesta, M., A. D'Amico, G. Pavan, A. Drougas, A. Komnenou, and N. Portunato. 2006. A review of Cuvier's beaked whale strandings in the Mediterranean Sea. *Journal of Cetacean Research and Management* 7(3): 251–261.

Pontoppidan, E. 1755. The natural history of Norway . . . In two parts. Translated from the Danish original of the Right Rev[d]. Erich Pontoppidan, . . . Illustrated with copper plates and a general map of Norway. A. Linde, London.

Praderi, R. 1972. Notas sobre un ejemplar de *Mesoplodon layardi* (Gray) (Cetacea, Hyperoodontidae), de la costa Atlantica de Uruguay. *Contributions of the Zoological Museum of Natural History, Montevideo* 10(37): 1–7.

Ralls, K. 1976. Mammals in which females are larger than males. *Quarterly Review of Biology* 51: 245–276.

Ralls, K., and R. L. Brownell. 1991. A whale of a new species. *Nature* 350: 560.

Rankin, J. J. 1953. First record of the rare beaked whale *Mesoplodon europaeus* Gervais from the West Indies. *Nature* 172: 873.

———. 1955. A rare whale in tropical seas (*Mesoplodon europaeus*). *Everglades Natural History* 3(1): 24–31.

———. 1956. The structure of the skull of the beaked whale *Mesoplodon gervaisi* Deslongchamps. *Journal of Morphology* 99(2): 329–358.

———. 1961. The bursa ovaries of the beaked whale *Mesoplodon gervaisi* Deslongchamps. *Anatomical Record* 139(3): 179–186.

Rankin, S., and J. Barlow. 2007. Sounds recorded in the presence of Blainville's beaked whale, *Mesoplodon densirostris*, near Hawai'i. *Journal of the Acoustical Society of America* 122(1): 42–45.

Rankin, S., S. Baumann-Pickering, T. Yack, and J. Barlow. 2011. Description of sounds recorded from Longman's beaked whale, *Indopacetus pacificus*. *Journal of the Acoustical Society of America* 130: EL339–EL344.

Raven, H. C. 1937. Notes on the taxonomy and osteology of two species of *Mesoplodon* (*M. europaeus* Gervais and *M. mirus* True). *American Museum Novitates* 905: 1–30.

———. 1934. Beaked Whales. *Natural History* 34(5): 501.

———. 1942. On the structure of *Mesoplodon densirostris*, a rare beaked whale. *Bulletin of the American Museum of Natural History* 80: 23–50.

Raven, H. C., and W. K. Gregory. 1933. The spermaceti organ and nasal passages of the sperm whale (*Physeter catodon*) and other odontocetes. *American Museum Novitates* 677: 1–17.

Red Data Book. 2008. Northern bottlenose whale, *Hyperoodon planifrons*. Code 11.08.5.V. IUCN, Morges, Switzerland.

Reeves, R. R., B. S. Stewart, P. J. Clapham, and J. A. Powell. 2002. *Guide to the Marine Mammals of the World*. National Audubon Society. Alfred A. Knopf.

Reeves, R. R., and F. A. Ulmer. 1976. New Jersey's medium-sized whales. *New Jersey Audubon* 11(12): 8–18.

Reidenberg, J. S. 2009. Anatomical adaptations of aquatic mammals. *Anatomical Record* 290: 507–513.

Reidenberg, J. S., and J. T. Laitman. 1994. Anatomy of the hyoid apparatus in Odontoceti (toothed whales): Specializations in their skeleton and musculature compared with those of terrestrial mammals. *Anatomical Record* 240: 598–624.

Rendell, L. 2012. Sperm whale communications and culture. *Whalewatcher* 41(1): 21–27.

Reyes J. C., J. G. Mead, and K. Van Waerbeek. 1991. A new species of beaked whale *Mesoplodon peruvianus* sp. n. (Cetacea: Ziphiidae) from Peru. *Marine Mammal Science* 7(1): 1–24.

Reyes, J. C., K. Van Waerbeek, J. C. Cardeñas and J. L. Yañez. 1995. *Mesoplodon bahamondi* sp. n. (Cetacea, Ziphiidae), a new living beaked whale from the Juan Fernández Archipelago, Chile. *Boletin del Museo Nacional de Historia Natural, Chile* 45: 31–44.

Rice, D. W. 1963. Progress report on biological studies of the larger cetaceans in the waters off California. *Norsk Hvalfangst-tidende* 52(7): 181–187.

———. 1977. A list of the marine mammals of the world. *NOAA Technical Report NMFS SSRF* 711: iii + 15 pp.

———. 1978. Beaked whales. In D. Haley, ed., *Marine Mammals of Eastern North Pacific and Arctic Waters*, pp. 88–95. Pacific Search Press.

———. 1998. *Marine Mammals of the World*. The Society for Marine Mammalogy.

———. 1989. Sperm whale. In S. H. Ridgway and R. Harrison, eds., *Handbook of Marine Mammals. Vol. 4: River Dolphins and the Larger Toothed Whales*, pp. 177–234. Academic Press.

Richards, L. P. 1952. Cuvier's beaked whale from Hawaii. *Journal of Mammalogy* 33(2): 355.

Ridgway, S. H. 1999. An illustration of Norris's acoustic window. *Marine Mammal Science* 15(4): 926–930.

Ritter, F., and B. Brederlau. 1999. Behavioural observations of dense-beaked whales (*Mesoplodon densirostris*) off La Gomera, Canary Islands (1995–1997). *Aquatic Mammals* 25(1): 55–61.

Robineau, D., and M. Vely. 1993. Stranding of a specimen of Gervais's beaked whale (*Mesoplodon europaeus*) on the coast of West Africa (Mauritania). *Marine Mammal Science* 19(4): 438–440.

Robinson, K. P., and C. D. MacLeod. 2009. First stranding report of a Cuvier's beaked whale (*Ziphius cavirostris*) in the Moray Firth in north-east Scotland. *Marine Biodiversity Records* 2: 1–3.

Robson, F. D. 1975. On vestigial and normal teeth in the scamper-down beaked whale. *Tuatara* 21(3): 1–3.

Roest, A. I. 1964. *Physeter* and *Mesoplodon* strandings on the central California coast. *Journal of Mammalogy* 45(1): 129–136.

Roest, A. I., M. Storm, and P. C. Dumas. 1953. Cuvier's beaked whale (*Ziphius cavirostris*) from Oregon. *Journal of Mammalogy* 34(2): 251–252.

Rommel, S. A., A. M. Costidis, A. Fernández, P. D. Jepson, D. A. Pabst, W. A. McLellan, D. S. Houser, T. W. Cranford, A. L. van Helden, D. M. Allen, and N. B. Barros. 2006. Elements of beaked whale anatomy and diving physiology and some hypothetical causes of sonar-related stranding. *Journal of Cetacean Research and Management* 7(3): 189–209.

Ross, G. J. B. 1969. Evidence for a southern breeding population of True's beaked whale. *Nature* 222: 585.

———. 1970. The occurrence of Hector's beaked whale *Mesoplodon hectori* (Gray) in South African waters. *Annals of the Cape Province Museum (Natural History)* 8(13): 195–204.

Ross, G. J. B., A. N. Baker, P. B. Best, and J. G. Mead. 1988. A review of colour patterns and their ontogenetic variation in beaked whales (Ziphiidae, Cetacea). *Scientific Reports of the International Whaling Commission* SC/40/SM8: 1–64.

Ross, G. J. B., A. N. Baker, R. N. P. Goodall, A. A. Lichter, and J. G. Mead. 1988. The distribution of beaked whales in the Southern Hemisphere. *Scientific Reports of the International Whaling Commission* SC/40/SM20: 1–35.

Ryder, J. A. 1887. On the development of the cetacea, together with a consideration of the probable homologies of the flukes of cetaceans and sirenians. *Report of the U. S. Fish Commission* 1885: 427–488.

Santos, M. B., V. Martin, M. Arbelo, A. Fernández, and G. J. Pierce. 2007. Insights into the diet of beaked whales from the atypical mass stranding in the Canary Islands in September 2002. *Journal of the Marine Biological Association of the U.K.* 87: 243–251.

Scheffer, V. B. 1949. Notes on three beaked whales from the Aleutian Islands. *Pacific Science* 3: 353.

Scheffer, V. B., and J. B. Slipp. 1948. The whales and dolphins of Washington State. *American Midland Naturalist* 39(2): 257–337.

Schevill, W. E. 1987. Reply to Holthuis—the scientific name of the sperm whale. *Marine Mammal Science* 3(1): 89–90.

Scholander, P. F. 1940. Experimental investigations on the respiratory function in diving mammals and birds. *Hvalradets Skrifter* 22(l): 1–131.

Schorr, G. S., R. W. Baird, M. B. Hanson, D. L. Webster, D. J. McSweeney, and R. D. Andrews. 2009. Movements of satellite-tagged Blainville's beaked whale off the island of Hawai'i. *Endangered Species Research* 10: 203–213.

Schorr, G. S., E. Falcone, et al. 2011. The bar is really noisy, but the food must be good: High site fidelity and dive behavior of Cuvier's beaked whales (*Ziphius cavirostris*) on an anti-submarine warfare range. *19th Biennial Conference on the Biology of Marine Mammals.* Tampa, FL. *Abstracts*, p. 265.

Sekiguchi, K., N. T. W. Klages, and P. B. Best. 1996. The diet of strap-toothed whales (*Mesoplodon layardii). Journal of Zoology* 239(3): 453–463.

Silverman, H. B., and M. J. Dunbar. 1980. Aggressive tusk use by the narwhal (*Monodon monoceros*). *Nature* 284: 57–58.

Simmonds, M. P., and L. F. Lopez-Jurado. 1991. Whales and the military. *Nature* 35 (6326): 448.

Simpson, G. G. 1945. The principles of classification and the classification of mammals. *Bulletin of the American Museum of Natural History* 85: 1–350.

Slijper, E. J. 1962. *Whales.* Cornell University Press.

Slip, D., G. J. Moore, and K. Green. 1995. Stomach contents of a southern bottlenose whale *Hyperoodon planifrons*, stranded at Heard Island. *Marine Mammal Science* 11(4): 575–584.

Slipp, J. W., and F. Wilke. 1953. The beaked whale *Berardius* on the Washington coast. *Journal of Mammalogy* 34(1): 105–113.

Smith, M. S. R. 1965. Fourth known individual of beaked whale genus *Tasmacetus. Mammalia* 29: 618–620.

Smithsonian Cetacean Distributional Database. http://collections.nmnh.si.edu/search/mammals/.

Sorensen, J. H. 1940. *Tasmacetus shepherdi*: History and description of specimens cast ashore at Mason's Bay, Stewart Island, in February, 1933. *Transactions of the Royal Society of New Zealand* 70: 200–204.

Southall, B. L., T. Rowles, F. Gulland, R. W. Baird, and P. D. Jepson. 2013. Final Report of the Independent Scientific Review Panel (ISRP) investigating potential contributing factors to a 2008 mass stranding of melon-headed whales (*Peponocephala electra*) in Antsohihy, Madagascar. WCS, IFAW, NOAA, IWC. Pp 1–75.

Southwell, T. 1883. On the beaked whale (*Hyperoodon rostratus*). *Transactions of the Norfolk Norwich Naturalist Society* 3: 476–481.

Sowerby, J. 1804. *The British Miscellany; or Coloured Figures of New, Rare or Little Known Animal Subjects, Many Not Before Ascertained to Be Inhabitants of the British Isles.* London.

Stejneger, L. 1883. Contributions to the history of the Commander Islands: Notes on the natural history, including descriptions of new cetaceans. *Proceedings of the U.S. National Museum* 6: 58–89.

Stephen, A. C. 1931. True's beaked whale (*Mesoplodon mirus*) new to the Scottish fauna. *Scottish Naturalist* 1931: 37–39.

Talbot, F. H. 1960. True's beaked whale from the south-east coast of South Africa. *Nature* 186: 406.

Taylor, B., J. Barlow, R. Pitman, L. Ballance, T. Klinger, D. DeMaster, J. Hildebrand, J. Urban, D. Placios, and J. Mead. 2004. A call for research to assess risk of acoustic impact on beaked whale populations. *Scientific Report of the International Whaling Commission* SC/56/E36.

Taylor, M. A. 1986. Stunning whales and deaf squids. *Nature* 323: 298–299.

Taylor, R. J. F. 1957. An unusual record of three species of whale being restricted to pools in Antarctic ice. *Proceedings of the Zoological Society of London* 29: 325–331.
Thewissen, J. G. M. 1994. Phylogenetic aspects of cetacean origins: Amorphological perspective. *Journal of Mammalian Evolution* 2: 157–184.
———. 1998. Cetacean origins: Evolutionary turmoil during the invasion of the oceans. In J. G. M. Thewissen, ed., *The Emergence of Whales,* pp. 451–464. Plenum.
Thewissen, J. G. M., and D. P. Domning. 1992. The role of phenacodontids in the origin of the modern orders of ungulate mammals. *Journal of Vertebrate Paleontology* 12(4): 494–504.
Thewissen, J. G. M., and F. E. Fish. 1997. Locomotor evolution in the earliest cetaceans: Functional model, modern analogues, and paleontological evidence. *Paleobiology* 23: 482–490.
Thewissen, J. G. M., and S. T. Hussain. 1998. Systematic review of the Pakicetidae, early and middle Eocene cetacea (Mammalia) from Pakistan and India. *Bulletin of the Carnegie Museum of Natural History* 34: 220–238.
Thewissen, J. G. M., S. I. Madar, and S. T. Hussain. 1996. *Ambulocetus natans,* an Eocene cetacean (Mammalia) from Pakistan. *Courier Forschungs-Institut Seckenberg* 191: 1–86.
Thompson, D. W. 1919. On whales landed at the Scottish whaling stations, especially during the years 1908–1914. *Scottish Naturalist* 85: 1–16.
Thompson, K., C. S. Baker, A. van Helden, S. Patel, C. Millar, and R. Constantine. 2012. The world's rarest whale. *Current Biology* 22(21): 905–906.
Tietz, R. M. 1966. The southern bottlenose whale, *Hyperoodon planifrons,* from Humewood, Port Elizabeth. *Annals of the Cape Province* 5: 101–107.
Tomilin, A. G. 1967[1957]. *Zveri SSSR i prilezhashchikh stran [Mammals of the U.S.S.R. and adjacent countries], volume 9, Kitoobraznye [Cetacea]." Akademi Nauk SSSR, Moskva,* 756 pp. [Translated from Russian by the Israel Program for Scientific Translations, 1967, IPST Cat. No. 1124, pp. xxi + 717; Published pursuant to an agreement with the Smithsonian Institution and the National Science Foundation; available from the Clearinghouse for Federal Scientific and Technical Information, TT 65-50086].
Tønnessen, J. N., and A. O. Johnsen. 1982. *The History of Modern Whaling.* C. Hurst and Company.
Tougaard, J., P. T. Madsen, and M. Wahlberg. 2008. Underwater noise fronm construction and operation of offshore wind farms. *Bioacoustics* 17(1–3): 143–146.
True, F. W. 1885. Contributions to the history of the Commander Islands. No. 5. Description of a new species of *Mesoplodon, M. stejnegeri,* obtained by Dr. Leonard Stejneger in Bering Island. *Proceedings of the U.S. National Museum* 8: 584–585.
———. 1910. An account of the beaked whales of the family Ziphiidae in the collection of the U.S. National Museum, with remarks on some specimens in other museums. *Bulletin of the U.S. National Museum* 73: 1–89.
———. 1913. Description of *Mesoplodon mirus,* a beaked whale recently discovered on the coast of North Carolina. *Proceedings of the U.S. National Museum* 45: 651–657.
Turner, W. 1872. On the occurrence of *Ziphius cavirostris* in the Shetland Seas, and a comparison of its skull with that of Sowerby's whale (*Mesoplodon sowerbyi*). *Transactions of the Royal Society of Edinburgh* 26: 769–780.

———. 1879. The form and structure of the teeth of *Mesoplodon layardii* and *Mesoplodon sowerbyi*. *Journal of Anatomical Physiology* 13: 465–480.

———. 1882. A specimen of Sowerby's whale (*Mesoplodon bidens*) captured in Shetland. *Journal of Anatomical Physiology* 16: 458–470.

———. 1886. On the occurrence of the bottle-nosed whale, (*Hyperoodon rostratus*) in the Scottish seas, with observations on its external characters. *Proceedings of the Royal Physiological Society of Edinburgh* 9: 25–47.

Tyack, P. L., M. Johnson, N. Aguilar de Soto, A. Sturlese, and P. T. Madsen. 2006. Extreme diving of beaked whales. *Journal of Experimental Biology* 209: 4238–4253.

Tyack, P. L., M. Johnson, P. T. Madsen, W. M. X. Zimmer, and N. Soto. 2008. How toothed whales echolocate to find and capture prey in the deep ocean. *Journal of the Acoustical Society of America* 123(5): 3104.

Tyack, P. L., W. M. X. Zimmer, D. Moretti, B. L. Southall, D. E. Claridge, J. W. Durban, C. W. Clark, A. D'Amico, N. Dimarzio, S. Jarvis, E. McCarthy, R. Morrisey, J. Ward, and I. L. Boyd. 2011. Beaked whales respond to simulated and actual navy sonar. *PLoS One* 6(3): e17009.

Uhen, M. D. 2009. Evolution of marine mammals: Back to the sea after 300 million years. *Anatomical Record* 290: 514–522.

Ulmer, F. A. 1941. *Mesoplodon mirus* in New Jersey with additional notes on the New Jersey *M. densirostris* and a list and key to the ziphioid whales of the Atlantic coast of North America. *Proceedings of the Academy of Natural Sciences of Philadelphia* 93: 107–122.

———. 1947. A second Florida record of *Mesoplodon europaeus*. *Journal of Mammalogy* 28(2): 184–185.

———. 1961. New Jersey's whales and dolphins. *New Jersey Nature News* 16(3): 80–93.

Urbán-Ramirez, J., and D. Aurioles-Gamboa. 1992. First record of the pygmy beaked whale *Mesoplodon peruvianus* in the North Pacific. *Marine Mammal Science* 8(4): 420–425.

Urbán-Ramirez, J., G. Cárdenas-Hinojosa, A. Gómez-Gallardo, Ú. Gonzales-Peral, W. del Toro-Orozco, and R. L. Brownell. 2007. Mass stranding of Baird's beaked whales at San Jose Island, Gulf of California, Mexico. *Latin American Journal of Aquatic Mammals* 6(1): 83–88.

Van Beneden, P. J., and P. Gervais 1880[1868–79]. *Ostéographie des cétacés, vivants et fossiles, comprenant la description et l'iconographie du squelette et du système dentaire de ces animaux, ainsi que des documents relatifs a leur histoire naturelle.* A. Bertrand, Paris.

van der Hoop, J., M. J. Moore, S. G. Barco, T. V. N. Cole, P.-Y. Doust, A. G. Henry, D. A. McAlpine, W. A. McLellan, T. A. Wimmer, and A. R. Solow. 2013. Assessment of management to mitigate anthropogenic effects on large whales. *Conservation Biology* 27(1): 121–135.

van Helden, A. L., A. N. Baker, M. L. Dalebout, J. C. Reyes, K. Van Waerbeek, and C. S. Baker. 2002. Resurrection of *Mesoplodon traversii* (Gray, 1874), senior synonym of *M. bahamondi* Reyes, Van Waerbeek, Cárdenas and Yáñez 1995 (Cetacea: Ziphiidae). *Marine Mammal Science* 18(3): 609–621.

Varona, L. S. 1964. Un craneo de *Ziphius cavirostris* del sur de Isla de Pinos. *Poeyana* ser. A(4): 1–3.

———. 1970. Morfologia externa y caracteres craneales de un macho adulto de *Mesoplodon europaeus* (Cetacea: Ziphiidae). *Poeyana* ser. A (69): 1–17.

Vrolik, W. 1848. "Natuur—en ontleedkundige beschouwing van den *Hyperoodon*." Natuurkundige Verhandlingen van de Hollandsche Maatschappij der Wetenschappen te Haarlem. 5th part (1st piece): i–128, pls. I–XV.

Wahlberg M., K. Beedholm, A. Heerfordt, and B. Møhl. 2011. Characteristics of biosonar signals from the northern bottlenose whale, *Hyperoodon ampullatus*. *Journal of the Acoustical Society of America* 130(5): 3077–3084.

Waite, E. R. 1922. Two ziphioid whales not previously recorded from South Australia. *Records of the South Australian Museum* 2(2): 209–214.

Walker, E. P. 1964. *Mammals of the World*. Johns Hopkins University Press, Baltimore.

Walker, W. A., and M. B. Hanson. 1999. Biological observations on Stejneger's beaked whale, *Mesoplodon stejnegeri*, from strandings on Adak Island, Alaska. *Marine Mammal Science* 15(4): 1314–1329.

Walker, W. A., J. G. Mead, and R. L. Brownell. 2002. Diets of Baird's beaked whales, *Berardius bairdii* in the southern Sea of Okhotsk and off the Pacific coast of Honshu, Japan. *Marine Mammal Science* 18(4): 902–919.

Waller, G. N. H. 2013. Note on James Sowerby and the discovery of Sowerby's beaked whale, *Mesoplodon bidens*. *Archives of Natural History* 40(2): 270–276.

Ward, J., S. Jarvis, D. Moretti, R. Morrisey, N. DiMarzio, M. Johnson, P. L. Tyack, L. Thomas, and T. Marques. 2011. Beaked whale (*Mesoplodon densirostris*) passive acoustic detection in increasing ambient noise. *Journal of the Acoustical Society of America* 129: 662.

Waring, G. T., T. Hamasaki, D. Sheehan, G. Wood, and S. Baker. 2001. Characterization of beaked whale (Ziphiidae) and sperm whale (*Physeter macrocephalus*) summer habitat in shelf-edge and deeper waters off the northeast U.S. *Marine Mammal Science* 17(4): 703–717.

Watkins, W. A. 1976. A probable sighting of a live *Tasmacetus shepherdi* in New Zealand waters. *Journal of Mammalogy* 57(2): 415.

———. 1977. Acoustic behavior of sperm whales. *Oceanus* 20(2): 50–58.

———. 1980. Acoustics and the behavior of sperm whales. In R.-G. Busnel and J. F. Fish, *Animal Sonar Systems*, pp. 283–290. Plenum.

Watkins, W. A., M. A. Daher, N. A. DiMarzio, A. Samuels, D. Wartzok, F. M. Fristrup, P. W. Howey, and R. R. Maiefski. 2002. Sperm whale dives tracked by radio tag telemetry. *Marine Mammal Science* 18(1): 55–78.

Watkins, W. A., and W. E. Schevill. 1976. Right whale feeding and baleen rattle. *Journal of Mammalogy* 57: 58–66.

———. 1977. Sperm whale codas. *Journal of the Acoustical Society of America* 62(6): 1485–1490.

———. 1979. Aerial observations of feeding behavior in four baleen whales: *Eubalaena glacialis*, *Balaenoptera borealis*, *Megaptera novaeangliae*, and *Baleanoptera physalus*. *Journal of Mammalogy* 60: 155–163.

———. 1982. Observations of right whales, *Eubaleana glacialis*, in Cape Cod waters. *Fishery Bulletin* 80(4): 875–880.

Watkins, W. A., and D. Wartzok. 1985. Sensory biophysics of marine mammals. *Marine Mammal Science* 1(3): 219–260.

Watwood, S. L., P. J. O. Miller, M. Johnson, P. T. Madsen, and P. L Tyack. 2006. Deep-diving foraging behavior of sperm whales (*Physeter macrocephalus*). *Journal of Animal Ecology* 75: 814–825.

Weilgart, L. S. 2007. A brief review of known effects of noise on marine mammals. *International Journal of Comparative Psychology* 20: 159–168.

Werth, A. J. 1990. Functional anatomy of the right whale tongue. *American Zoologist* 30: 21A.

———. 2001. How do mysticetes remove prey trapped in baleen? *Bulletin of the Museum of Comparative Zoology* 156(1): 189–203.

———. 2004a. A kinematic study of suction feeding and associated behaviors in the long-finned pilot whale, *Globicephala melas* (Traill). *Marine Mammal Science* 16(2): 299–314.

———. 2004b. Functional morphology of the sperm whale tongue, with reference to suction feeding. *Aquatic Mammals* 30: 405–418.

———. 2004c. Models of hydrodynamic flow in the bowhead whale filter feeding apparatus. *Journal of Experimental Biology* 207(20): 3569–3580.

———. 2006a. Mandibular and dental variation and the evolution of suction feeding in the Odontoceti. *Journal of Mammalogy* 87(3): 579–588.

———. 2006b. Odontocete suction feeding: Experimental analysis of water flow and head shape. *Journal of Morphology* 267: 1415–1428.

———. 2007. Adaptations of the cetacean hyolingual apparatus for aquatic feeding and thermoregulation. *Anatomical Record* 290: 546–568.

West, K. L., S. Sanchez, D. Rotstein, K. M. Robertson, S. Dennison, G. Levine, N. Davis, D. Schofield, C. W. Potter, and B. Jensen. 2013. A Longman's beaked whale (*Indopacetus pacificus*) strands in Maui, Hawaii, with first case of mobillivirus in the central Pacific. *Marine Mammal Science* 29(4): 767–776.

Whitehead, H. 1986. Call me gentle. *Natural History* 95(6): 4–11.

———. 1990. *Voyage to the Whales.* Chelsea Green.

———. 1995. The realm of the elusive sperm whale. *National Geographic* 188(5): 57–73.

———. 1996. Variation in the feeding success of sperm whales: Temporal scale, spatial scale and relationship to migrations. *Journal of Animal Ecology* 65: 429–438.

———. 2003. *Sperm Whales: Social Evolution in the Ocean.* University of Chicago Press.

———. 2009. Sperm whale *Physeter macrocephalus.* In W. F. Perrin, B. Würsig, and J. G. M. Thewissen, eds., *Encyclopedia of Marine Mammals,* pp. 1091–1098. Academic Press.

———. 2012. The whale that captured me. *Whalewatcher* 41(1): 2–8.

Whitehead, H., and L. Weilgart. 1990. Click rates from sperm whales. *Journal of the Acoustical Society of America* 87(4): 1798–1808.

Willett, C., and P. Cunningham. 1951. *The History of Underclothes.* Michael Joseph, London [Dover Publications, Inc., New York, 266 pp.].

Williams, D. 2016. *Why whales strand.* http://deafwhale.com/author/davidwilliams/.

Willis, P. M., and R. W. Baird. 1998. Sightings and strandings of beaked whales on the west coast of Canada. *Aquatic Mammals* 24: 21–25.

Wilson, D. E., and D. M. Reeder, eds. 2005. *Mammal Species of the World.* Third edition. Johns Hopkins University Press, Baltimore.

Wilson, E. O. 1992. *The Diversity of Life.* Harvard University Press.

Wilson, M., R. T. Hanlon, P. L. Tyack, and P. T. Madsen. 2007. Intense ultrasonic clicks from echolocating toothed whales do not elicit anti-predator responses or debilitate the squid *Loligo pealeii*. *Biological Letters* 3: 225–227.

Winn, H. E., P. J. Perkins, and L. Winn. 1970. Sounds and behavior of the northern bottle-nosed whale. *Proceedings of the Seventh Annual Conference on Biological Sonar and Diving Mammals 1970*, pp. 53–59. Stanford Research Institute.

Woodside, J. M., L. David, A. Frantzis, and S. K. Hooker. 2006. Gouge marks on deep-sea mud volcanoes in the eastern Mediterranean: Caused by Cuvier's beaked whales? *Deep Sea Research Part I* 53(11): 1762–1771.

Worthington, L. V., and W. E. Schevill. 1957. Underwater sounds heard from sperm whales. *Nature* 180: 291.

Wright, A. J., N. A. Aguilar Soto, A. L. Baldwin, M. Bateson, C. M. Beale, C. Clark, T. Deak, E. F. Edwards, A. F. and Ana Godinho, et al. 2007. Do marine mammals experience stress related to anthropgenic noise? *International Journal of Comparative Psychology* 20: 274–316.

Yamada, T. K. 1996. Tooth growth of Stejneger's beaked whale. *Nihonkai Cetology* 6: 1–5.

Yamada, T. K., Y. Tajima, A. Yatabe, B. M. Allen, and R. L. Brownell. 2012. Review of current knowledge on Hubbs's beaked whale, *Mesoplodon carlhubbsi*, from the seas around Japan and data from North America. Paper SC/62/SM17 presented to the IWC Scientific Committee (unpublished). 8 pp.

Zerbini, A. N., and E. R. Secchi. 2001. Occurrence of Hector's beaked whale, *Mesoplodon hectori*, in southern Brazil. *Aquatic Mammals* 27(2): 149–153.

Zimmer, W. M. X., J. Harwood, M. P. Johnson, P. L. Tyack, and P. T. Madsen. 2008. Passive acoustic detection of deep-diving beaked whales. *Journal of the Acoustical Society of America* 124(5): 2823–2832.

Zimmer, W. M. X., M. Johnson, P. T. Madsen, and P. L. Tyack. 2005. Echolocation clicks of a free-ranging Cuvier's beaked whale (*Ziphius cavirostris*). *Journal of the Acoustical Society of America* 117: 3919–3927.

Zimmer, W. M. X., and P. L. Tyack. 2007. Repetitive shallow dives pose decompression risk in deep-diving beaked whales. *Marine Mammal Science* 23(4): 888–925.

Zioupos, P., J. D. Currey, A. Casinos, and V. de Buffrénil. 1997. Mechanical properties of the rostrum of the whale *Mesoplodon densirostris*, a remarkably dense bony tissue. *Journal of Zoology* 341: 725–737.

Index

Note: Subentries (distribution, size, etc.) occur under the common names, not the scientific names. Page numbers in bold indicate species chapters.